Doubled Haploids: Technological Advances and Role In Crop Improvement

Zenu Jha • Satish B. Verulkar
Suprasanna Penna
Editors

Doubled Haploids: Technological Advances and Role In Crop Improvement

Springer

Editors
Zenu Jha
Department of Plant Molecular Biology & Biotech
Indira Gandhi Agricultural University
Raipur, Chhattisgarh, India

Satish B. Verulkar
Department of Plant Molecular Biology & Biotech
Indira Gandhi Agricultural University
Raipur, Chhattisgarh, India

Suprasanna Penna
Amity Center for Nuclear Biotechnology
Amity Institute of Biotechnology
Amity University of Maharashtra
Mumbai, Maharashtra, India

ISBN 978-981-96-2338-9   ISBN 978-981-96-2339-6 (eBook)
https://doi.org/10.1007/978-981-96-2339-6

This Springer imprint is published by the registered company Springer Nature Singapore Pte Ltd.
The registered company address is: 152 Beach Road, #21-01/04 Gateway East, Singapore 189721, Singapore

Foreword

Globally, food security has been the prime focus of all the crop breeding and improvement programmes. Challenges of meeting food supply and food security have become a concern against the backdrop of increasing human population, climate change and environmental extremities. In this context, although conventional breeding has resulted in varietal development, implementation innovative technologies in breeding has become essential for the development of new improved crop varieties. Biotechnological and genomics advances have been successfully developed to accelerate plant breeding methods. Doubled haploid (DH) stands at the forefront of crop improvement as a unique technology for basic and applied genetic research and offers to produce pure, homozygous lines needed for large-scale hybrid seed production. Relying initially on the in vivo, in vitro methods for the production of haploid plants, technological advances have now been evolved through genetic engineering and gene editing methods.

This book 'Doubled Haploids: Technological Advances and Role In Crop Improvement' has covered the concepts and different facets of haploids including successful examples. The technological know-how presented is very useful for rapid generation of new inbred lines and hybrids, mapping of genomic regions, identification of markers for haploid selection and genome editing approaches. This book brings in excellent information for plant scientists for understanding doubled haploids and to engage in developing high yielding, climate resilient crops.

Doubled Haploids is an emerging area of plant biology and biotechnology with considerable scope in crop breeding. The editors of this illustrious book have done a commendable work and congratulations to the editors for compiling such a book. I have no doubt that this will become a valuable resource for students, plant biologists, genomics researchers and breeders.

International Chair—Agriculture & Food Security
Murdoch University,
Murdoch, WA, Australia

Rajeev K. Varshney

Preface

Current and future demands of food have become a formidable challenge in the backdrop of looming threat from climate change, land degradation and increasing population pressure. This situation and challenges therein necessitate that we embrace new technologies and integrate them into food production systems. Crop improvement options using conventional and biotechnological breeding have been demonstrated to address these food security challenges and enable the development of improved crop varieties. Among the biotechnological solutions, development of doubled haploids is a novel technology which has undergone a fascinating evolution over the years, from in vivo to in vitro and the present, molecular method of haploid plant generation, and doubled haploids production.

Doubled haploid (DH) technology stands at the forefront of crop improvement, enabling faster, more precise breeding. It has emerged as a transformative advancement in crop breeding as it allows the rapid development of stable inbred lines with 100% homozygosity and significantly shortens the breeding process, making it commercially viable. The haploids, discovered as early as 1921 by Bergner in *Datura stramonium*, by Indian scientists Guha and Maheswari in 1964 reported direct development of haploid embryos and plantlets from microspores of *Datura innoxia* through anther cultures. Subsequently, researchers achieved the first full-fledged haploid plants from *Nicotiana tabacum*. Despite such remarkable advantages, the technology has seen some technical breakthroughs during 1980s and 1990s but the technological and translational resurgence during the past decade is now poised to making it as a valuable tool for plant breeding. Several new methodological advances have been made in different crop plants.

Through this book, efforts have been made to collate the up-to-date technological happenings in the field of doubled haploids. We have comprehended chapters in the area of DH production ranging from the historical developments, technological breakthroughs to the recent advancements and success stories. In the Chap. 1, we have described how haploids, once a scientific curiosity, now can contribute significantly to crop enhancement, ensuring food security. This chapter will take readers on a journey where early researchers focused on understanding haploid formation and its potential applications while the advancements mainly focused on various haploid induction technologies. The following Chap. 2 covers various methods for in vitro culturing of haploids. In vitro haploid production is a fast and reliable approach for crop improvement. In recent years, in vitro methods have revolutionized haploid production, offering rapid and efficient ways to achieve homozygosity. Next Chap. 3 summarizes the methodologies pertaining to creation of maternal haploids and their application in crop breeding. The Chap. 4 presents the resurgence in the advancement of haploid technology through a synthesis of recent techniques and advancements made in haploid technology to accelerate haploid production in crop plants. Taking cues from maize and rice haploids, the next Chap. 5 described the successful outcomes of research advances, while summarizing how both maize and rice haploids contribute significantly to crop improvement, genetic research and accelerated breeding processes. In the last Chap. 6, we have made an effort to present a current and fair account of success stories involving doubled haploids in different plant species.

It has been our endeavour to present a book having both the theoretical and practical aspects of doubled haploids technology. We hope that the book will be helpful to graduate and post-graduate students, teachers, researchers working in the areas of plant breeding, crop improvement, crop science, biotechnology and genetic engineering. We feel the information will add more to their understanding on how to apply this technique to a variety of crops that are vital to food security and sustainability. We would like to acknowledge authors for their tremendous work in contributing to this book. We sincerely thank the Springer Nature and its editorial team for the timely support in completing the publication of this book.

Raipur, Chhattisgarh, India — Zenu Jha
Raipur, Chhattisgarh, India — Satish B. Verulkar
Mumbai, Maharashtra, India — Suprasanna Penna

Contents

Editors and Contributors

About the Editors

Zenu Jha, a distinguished Ph.D. holder in Biotechnology, currently holds the position of Professor within the Department of Plant Molecular Biology and Biotechnology at Indira Gandhi Krishi Vishwavidyalaya, located in Raipur, Chhattisgarh, India. With an illustrious career spanning two decades in both teaching and research, Dr. Jha specializes in plant tissue culture, exploring its diverse applications such as anther culture, extraction of novel metabolites from plants and commercial production techniques. Notably, her pioneering work led to the development of a double haploid rice line characterized by high yield and robust resistance to multiple diseases. Throughout her career, Dr. Jha has demonstrated a keen commitment to innovation and intellectual property, having filed two patents for her groundbreaking research. She has successfully led numerous research projects, supported by prestigious funding bodies including the Department of Biotechnology (DBT), National Horticulture Mission (NHM), National Biodiversity Mission (NBM), Rashtriya Krishi Vikas Yojana (RKVY) and State Research Councils. Dr. Jha's contributions to academia extend beyond her laboratory work, with over 40+ research papers and book chapters to her credit, reflecting her dedication to advancing scientific knowledge. Her exemplary achievements have earned her recognition and accolades, including the esteemed Young Scientist Award, Pragati Sheel Mahila Samman, the Best Worker Award from Indira Gandhi Krishi Vishwavidyalaya, India, etc.

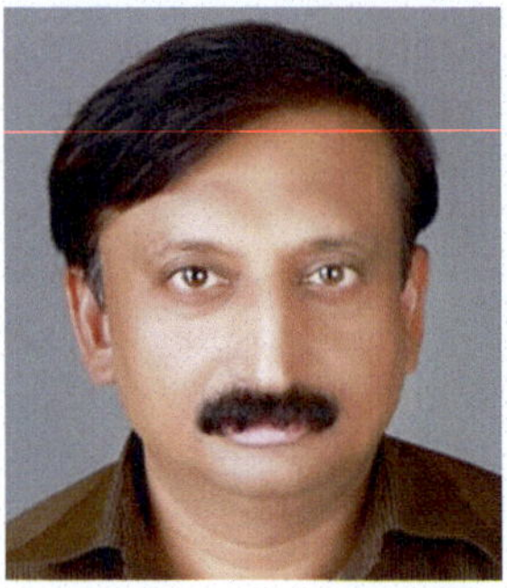

Satish B. Verulkar, a distinguished scholar holding a Ph.D. in Plant Breeding, currently serves as a Professor within the Department of Plant Molecular Biology and Biotechnology at Indira Gandhi Krishi Vishwavidyalaya in Raipur, Chhattisgarh, India. Renowned globally as a leading figure in drought breeding, Dr. Verulkar has an impressive career spanning over 32 years, dedicated to both teaching and research. His groundbreaking work has resulted in the development of nine distinct rice varieties, solidifying his reputation as an expert in Marker-Assisted Selection (MAS)-based breeding and DNA fingerprinting techniques across crop species. Dr. Verulkar's research endeavours have been generously supported by esteemed national and international agencies, including the International Rice Research Institute (IRRI), Stress-Tolerant Rice for Africa and South Asia (STRASA), the Department of Biotechnology (DBT) and the Rockefeller Foundation. Dr. Verulkar has authored over 80+ research papers in esteemed journals and contributed to various book chapters, showcasing his significant contributions to advancing agricultural science. His outstanding achievements have garnered widespread recognition, with prestigious awards and honours bestowed upon him by institutions such as the Rockefeller Foundation USA, STRASA IRRI Philippines, as well as accolades like the Outstanding Teacher Award and the Best Worker Award from Indira Gandhi Krishi Vishwavidyalaya, India.

Suprasanna Penna, Ph.D. in Genetics, is a Professor and Director at the Amity Institute of Nuclear Biotechnology, Amity University of Maharashtra, Mumbai, India. He was formerly with the Bhabha Atomic Research Centre, Mumbai, as the Head of Nuclear Agriculture and Biotechnology Division. He has served as an Expert consultant on IAEA missions on mutation breeding in crop plants. He has made significant contributions to plant science covering plant biotechnology, mutation breeding, stress tolerance and radiation-depolymerized oligochitosan, which have been successfully used in enhancing plant productivity in crop plants. He has conducted research on plant abiotic stress tolerance in crop plants and salinity-adaptive mechanisms in halophytes. He is on the editorial board of several international journals and has published over 390 publications and edited several books on plant mutation breeding, salinity tolerance, plant genetic diversity, non-coding RNAs and plant-metal interactions.

Contributors

Garima Diwan Department of Genetics and Plant Breeding, Indira Gandhi Krishi Vishwavidyalaya, Raipur, Chhattisgarh, India

Mahima Dubey VNR Seeds Private Limited, Raipur, Chhattisgarh, India

Zenu Jha Department of Plant Molecular Biology and Biotechnology, Indira Gandhi Krishi Vishwavidyalaya, Raipur, Chhattisgarh, India

Department of Genetics and Plant Breeding, Indira Gandhi Krishi Vishwavidyalaya, Raipur, Chhattisgarh, India

Tanveen Kaur Roy J. Carver Department of Biochemistry, Biophysics and Molecular Biology, Iowa State University of Science and Technology, Iowa, USA

Kanushree Nandedkar Department of Plant Molecular Biology and Biotechnology, Indira Gandhi Krishi Vishwavidyalaya, Raipur, Chhattisgarh, India

Nirmala Bharti Patel College of Agriculture & Research Station (CARS), Mahasamund, IGKV, Raipur, India

Suprasanna Penna Amity Center for Nuclear Biotechnology, Amity Institute of Biotechnology, Amity University Maharashtra, Mumbai, India

Suman Rawte College of Agriculture & Research Station (CARS), Gariaband, IGKV, Raipur, Chhattisgarh, India

Department of Genetics and Plant Breeding, Indira Gandhi Krishi Vishwavidyalaya, Raipur, Chhattisgarh, India

Vishal Sharma Biotechnology Division, CSIR-Institute of Himalayan Bioresource Technology, Palampur, Himachal Pradesh, India

Satish B. Verulkar Department of Plant Molecular Biology and Biotechnology, Indira Gandhi Krishi Vishwavidyalaya, Raipur, Chhattisgarh, India

Haploids: Then and Now

1

Garima Diwan, Suman Rawte, and Zenu Jha

Abstract

Possessing a single set of chromosomes, haploid organisms have long intrigued scientists due to their unique genetic characteristics and evolutionary significance. Historically, the field began with pioneering experiments that led to the generation of haploid plants, offering a unique platform for studying plant genetics. The advent of in vitro culture techniques in the mid-twentieth century marked a significant milestone, enabling efficient production of haploid plants from various tissues. For haploid generation two different approaches can be used: in vitro and in vivo (in seed). Through the in vitro development of gametophytes (microspores and megaspores) or the floral tissues or organs that surround them (anthers, ovaries, or ovules), in vivo techniques include wide crossing, pollen irradiation and in some cases the use of genetically defective haploid inducer lines. Moreover, the application of molecular markers and genomics tools has expedited the identification and selection of haploid plants carrying desirable traits. Gene editing technology has further facilitated the development of haploid plants with targeted genetic modifications revolutionizing the field of functional genomics and trait improvement. By synthesizing historical perspectives with contemporary insights, this chapter provides a comprehensive overview of the enduring significance of haploidy in genetics, genomics, and biotechnology highlighting the remarkable progress in haploidy.

G. Diwan · Z. Jha (✉)
Indira Gandhi Krishi Vishwavidyalaya, Raipur, Chhattisgarh, India

S. Rawte
College of Agriculture & Research Station (CARS), Gariaband, IGKV, Raipur, Chhattisgarh, India

Z. Jha et al. (eds.), *Doubled Haploids: Technological Advances and Role In Crop Improvement*, https://doi.org/10.1007/978-981-96-2339-6_1

Keywords

Haploid · Historical background · Methods and techniques · Recent advances

1.1 Introduction

Haploid plant production is a fundamental technique in plant breeding and has undergone a fascinating evolution over the years. A haploid cell is a cell that contains a single set of chromosomes. The doubled haploid (DH) process is initiated with the induction of haploid (1N) plants. Homologous chromosomal pairing is necessary for successful meiosis because haploid plants are usually sterile. Floral cell lineages need to go through doubling (DNA replication without mitosis) in order for haploids to set seed. This process can be stimulated by administering a microtubule polymerization inhibitor such as colchicine (Kelliher et al. 2016). Most sexually reproducing organisms are diploid, meaning they have two sets of chromosomes, one from each parent. Haploid organisms are relatively rare, but they can be found in both plants and animals. Some examples of haploid organisms include pollen grain, egg cells, sperm cells, some fungi, some algae, and some insects. In nature, haploids arise as an abnormality when the haploid egg or a synergid forms an embryo without fertilization (Bhojwani et al. 2010). Having a gametophytic chromosome number, haploid plants are autonomous and sporophytic, deriving from a gametic cell found in the embryo sac or pollen grain. The haploid embryo can develop from a male gamete (androgenesis), an egg cell (gynogenesis), or a gametophytic cell other than an egg cell (apogamy). When pollen or anthers are cultivated in vitro, it can also come from the microspore nucleus before the first pollen grain mitosis. Known as monoploids, haploids descended from exclusively diploid species have a single genome as contrasted with haploids produced from polyploid species, which have many genomes (Jain et al. 1996).

Historically, haploidy was instrumental in shaping fundamental genetic theories such as Mendel's laws of inheritance, providing a foundational understanding of heredity. The first haploid plants were discovered in the early 1900s by German botanist Erwin Baur. Baur was studying the genetics of the Jimson weed, *Datura stramonium*, when he discovered a plant that had only one set of chromosomes. He called this plant a haploid. Haploids are found in a smaller number of animal species than plants. Some animals are naturally haploid, such as some insects and some nematodes. Since these early discoveries, scientists have learned a great deal about haploids. They have found that haploids are relatively common in both plants and animals. They have also found that haploids can be used to study a variety of genetic and biological phenomena.

Haploid plants were primarily obtained through laborious and often unpredictable methods, but over time advances in molecular biology, genomics, and biotechnology have transformed our understanding of haploidy. Plant breeders are interested in any methodological advancements as well as new haploidization principles because of the limitations of the current haploid technology. When using haploid technology in its most common format, plants grow directly from gametophytic (haploid) cells found in a range of explants that are grown in vitro such as pistils, ovules, anthers, or isolated immature pollen (usually microspores but less frequently early bicellular pollen). The regenerated plants are referred to as maternal or paternal haploids, depending on whether they receive their haploid genome from female or male gametophytes (embryo sac and pollen), respectively. Alternatively, uniparental genome elimination, which normally occurs during early zygotic embryogenesis, can result in the production of haploids through intra- or interspecific crossings. Another possibility is that parthenogenesis and apogamy, chromosome elimination, and somatic reduction have been documented in plants that have undergone genetic or hormonal manipulation (Jain et al. 1996). After pollination, haploidy may also develop if the pollen has been exposed to radiation, chemicals, or high temperatures. As a result, compared to standard practice, there is significant time and resource savings. Though there are several methods for producing haploid plants, viable haploid technology is either not yet available in many crop species, or it is only applicable to a limited number of genotypes and is extremely costly (Ishii et al. 2016; Kumlehn and Stein 2014; Shen et al. 2015; and rew. by Dwivedi et al. 2015). Even if these percentages seem low, a sizable number of viable hybrids and haploids can be grown in very small soil containers due to the enormous supply of tiny seeds and the early mortality of the majority of seedlings.

Haploids are used in the breeding of self-pollinated crops because of their ability to produce entirely homozygous lines in a single generation, which can be evaluated in the field. Haploids can be used for early generation testing of segregating populations because doubled haploids only have additive variance. Haploids can also be used in conventional breeding programs to increase selection reliability and program efficiency. The use of haploids in cross-pollinated crop breeding also relies on the rapid development of DH lines, which can be employed as inbred lines to create hybrid varieties. These essential hurdles can be overcome by haploids (Hermsen and Ramanna 1981).

The "doubled haploid" (DH) method is a useful tool for plant breeders since it allows them to quickly create homozygous seedlings. Thus, the genetically homogenous material produced by progenies of DH plants allows for boosting (Forster and Thomas 2005a, b; Wędzony et al. 2009; Gilles et al. 2017) breeding efficiency by enabling breeders to assess desirable features on genetically fixed material early in the breeding cycle. Several mechanisms are known to result in haploid plants. The two primary processes of the DH technology are (1) producing haploid embryos or plantlets through a haploid induction method and (2) restoring the diploidy of these plantlets through chromosome doubling (copy and paste). Stated differently, this technology involves the generation of several haploid plantlets that subsequently undergo complete genome duplication, resulting in a precisely homozygous plants (Forster and Thomas 2005a, b; Wędzony et al. 2009; Gilles et al. 2017; Maluszynski et al. 2003a, 2020, d; Nathanaël et al. 2020) (Table 1.1).

In recent years, technological innovations particularly in cell and molecular biology have revolutionized haploid plant production. Modern genomic techniques have unveiled intricate mechanisms governing haploid genomes, shedding light on

Table 1.1 Example of crop plants and vegetables in which haploid plantlet production is used in breeding, together with the main haploid induction system used

Crop name	Main method used to produce haploid embryos or plantlets	References
Maize	In planta haploid inducer line	Geiger and Gordillo (2009)
Wheat	Anther or microspore culture, interspecific cross	Maluszynski et al. (2003a, b, c), Touraev et al. (2009), Bajaj (1990), and Liu et al. (2020a, b)
Barley	Anther or microspore culture, interspecific cross, in planta haploid inducer line	Maluszynski et al. (2003a, b, c), Touraev et al. (2009), and Bajaj (1990)
Ryegrass	Anther or microspore culture	Maluszynski et al. (2003a, b, c) and Begheyn et al. (2016)
Triticale	Anther or microspore culture, interspecific cross	Maluszynski et al. (2003a, b, c) and Touraev et al. (2009)
Flax	Anther or microspore culture	Maluszynski et al. (2003a, b, c)
Rapeseed	Anther or microspore culture	Maluszynski et al. (2003a, b, c) and Touraev et al. (2009)
Melon	Irradiated pollen	Dong et al. (2016)
Cucumber	Ovary or ovule culture, irradiated pollen	Jain et al. (1996) and Dong et al. (2016)
Sugar beet	Ovary or ovule culture	Maluszynski et al. (2003a, b, c), Touraev et al. (2009), Bajaj (1990), and Jain et al. (1996)
Chili pepper and sweet pepper	Anther or microspore culture	Bajaj (1990)
Aubergine/eggplant	Anther or microspore culture	Jain et al. (1996)
Onion	Ovary or ovule culture	Maluszynski et al. (2003a, b, c), Touraev et al. (2009), and Jain et al. (1996)
Carrot	Anther or microspore culture	Touraev et al. (2009) and Bajaj (1990)
Asparagus	Anther or microspore culture	Maluszynski et al. (2003a, b, c)
Cauliflower, cabbage and broccoli	Anther or microspore culture	Maluszynski et al. (2003a, b, c) and Bajaj (1990)
Potato	Anther or microspore culture, in planta haploid inducer line	Maluszynski et al. (2003a, b, c), Touraev et al. (2009), and Jain et al. (1996)
Sorghum	In planta haploid inducer line	Hussain and Franks (2019) and Kloiber-Maitz et al. (2018)
Rice	Anther or microspore culture, in planta haploid inducer line	Maluszynski et al. (2003a, b, c), Touraev et al. (2009), Bajaj (1990), and Yao et al. (2018)
Soybean	Anther or microspore culture	Jain et al. (1996)
Loquat	Parthenogenesis (gamma radiation)	Blasco (2016)

their roles in disease susceptibility, adaptation, and evolution. These developments laid the foundation for wide-scale haploid breeding programs enhancing the speed and precision of crop improvement efforts. The introduction of in vitro culture methods in the mid-twentieth century marked a paradigm shift enabling the efficient production of haploid plants from diverse plant species. Advanced tissue culture protocols, coupled with hormonal manipulation and stress treatments, have significantly enhanced the efficiency of haploid embryo induction. Moreover, the integration of molecular markers, genomics, and bioinformatics tools has expedited the identification of haploid-inducing genes and facilitated the selection of haploid plants with desirable traits. The advent of genome editing technologies particularly CRISPR-Cas9 has further accelerated the precise modification of plant genomes, opening new avenues for functional genomics and trait improvement (Vijverberg et al. 2019). This chapter provides a detailed account of the historical milestones, including the development of classical and modern methods, such as anther and microspore culture, androgenesis, and gynogenesis. It explores the underlying physiological and molecular mechanisms governing haploid plant production, shedding light on the challenges and breakthroughs encountered by researchers. Additionally, we discuss the applications of haploid plants in crop improvement, emphasizing their roles in accelerating breeding programs, developing genetically modified organisms, and studying gene function.

Today there are a variety of reliable methods for producing haploid plants from a wide range of plant species. Haploid plant production is now a valuable tool for plant breeders and geneticists who are working to improve crops and develop new knowledge about plant biology. Haploid plants are also used in genetic research. For example, they can be used to identify and map genes that are responsible for important traits. Haploid plants can also be used to produce new mutant lines that can be used to study gene function. New technologies are being developed that will make it even more efficient and effective to produce haploid plants.

Doubled haploid (DH) production allows the encapsulation of haploid genomes within inbred lines in a single generation, in contrast to traditional plant breeding methods that necessitate numerous generations of self-pollination and/or backcrossing (Jacquier et al. 2020). Through in vitro tissue culture, in vivo interspecific and intraspecific hybridization with inducer lines, or manipulation of CENH3 centromeric histone proteins, haploid gametophytes can be induced to develop as DH sporophytes (Coe 1959; Seguí-Simarro and Nuez 2007; Ravi and Chan 2010; rew. by Zhang et al. 2020).

Furthermore, the potential of haploid technologies in addressing global agricultural challenges by synthesizing historical perspectives with contemporary insights offers a comprehensive understanding of haploid plant production, emphasizing its pivotal role in shaping modern agriculture and biotechnology. The integration of traditional wisdom with cutting-edge innovations exemplifies the dynamic nature of this field, underscoring its significance in sustainable crop improvement and the pursuit of food security.

Having the ability to produce haploids and then induce the entire genome duplication has made it possible to speed up crop breeding considerably. Breeders benefit greatly from the ensuing doubled haploids since all loci are simultaneously genetically fixed within a single generational step. By doing this, the laborious traditional need for substantial selfing or backcrossing to produce breeding inbred lines is evaded. Breeders can assess as many traits as necessary using genetically homogeneous material and extraordinarily selected offspring at a very early stage of the breeding cycle, as a result of the ability to quickly and identically reproduce the genotype of a doubled haploid line. This chapter comprehensively discusses the historical developments and recent advancements in haploid plant production techniques spanning from early discoveries to contemporary biotechnological innovations.

1.1.1 Haploidy in Evolution

Haploid is an organism that looks like a sporophyte but has the chromosome complement of a reduced gamete. Haploids can be produced in vivo or in vitro (Hermsen and Ramanna 1981). The genetic material of all eukaryotes is organized in the form of distinct chromosomes that reside within the cell nucleus. Different eukaryotes have strikingly diverse numbers of chromosomal copies or ploidies (n). Aside from diploidy (2n), polyploidy (having more than two chromosomal sets per nucleus) is a feature of various species (Comai 2005). Polyploidy is highly frequent and variable in plants (Adams and Wendel 2005; Soltis et al. 2015), as exemplified by the triploid (3n) banana (genus *Musa*) (Heslop-Harrison and Schwarzacher 2007), tetraploid (4n) cotton (genus *Gossypium*) (Wendel and Richard 2003), hexaploid (6n) wheat (genus *Triticum*) (Shewry 2009), and octoploid (8n) and decaploid (10n) strawberry species (genus *Fragaria*) (Liston et al. 2014). Polyploidy arises from an increase in the number of complete sets of chromosomes. Conversely, the reverse transition, i.e., from diploidy to haploidy (1n), requires that the number of chromosomal copies is reduced by half, usually through meiosis. Remarkably, numerous organisms across eukaryotic phylogeny are viable as haploids, carrying only one set of chromosomes in at least some part of their life cycle and/or one of the sexes. Thus the study of "haploids then and now" provides new opportunities for addressing basic questions on ploidy requirements in plant development and evolution, particularly in crops.

1.1.1.1 Historical Background of Haploids in Plants

Throughout the nineteenth, twentieth and twnety-first centuries, the study and manipulation of haploid cells in plants have contributed significantly to the fields of genetics, reproduction, and biotechnology leading to numerous scientific advancements and applications.

Discovery of the First Natural Haploid Plant

Since the discovery of the first "natural" haploid plant in *Datura stramonium* in 1921 by A.D. Bergner (Blakeslee et al. 1922), the second found in *Nicotiana tabacum* was likewise in solanaceous crop (Clausen and Mann 1924). The discovery of haploidy reported by Blakeslee significantly increased the interest in growing haploid plants. Botanists and geneticists have been acutely aware of the possibilities of haploids; it is interesting that approximately 40 years later the same genus was also the subject of the discovery of anther culture for haploid production.

The idea of pure lines had been established by Johannsen (1903, 1909) by the 1920s, and the gene theory of heredity was also well-known. Similarly, Muller's (1927) research on mutations and the potential for broad induction by outside agents had also gained recognition. Therefore, the benefit of haploids was evident and drew the curiosity of numerous well-known geneticists and plant breeders. Haploid had actually been discovered in a wide range of plants including significant crops like cotton, rice wheat, potato, tobacco, maize, and barley. Comprehensive assessments of haploids known from nature as well as those whose origin was thought to be human were compiled by Kimber and Riley (1963), Katayama and Nei (1964), and Nei (1963).

Early Discoveries and Foundations: Nineteenth Century

An early attempt at haploid induction by experimentation took place between 1920 and 1950. According to Chase (1969), the first haploid angiosperm for which cytological proof was obtained was discovered in 1921 by Dorothy Bergner in *Datura stramonium* (Blakeslee et al. 1922). The two haploid individuals were among a number of plants of abnormal appearance produced in an attempt to induce chromosomal irregularities by the application of cold as a stimulus. The first confirmed haploid from a cereal species was reported by Gaines and Aase (1926) from *Triticum compactum* var. *humboldtii*, a winter variety of the Pacific Northwest. In the years that followed, haploids were reported in a variety of species. Later on, haploids were discovered in numerous additional species like *Hordeum vulgare* (Johansen 1934), *Oryza sativa* (Morinaga and Fvkushima 1931), and *Triticum compactum*. It was reported that haploids were found in Compositae, Solanaceae, and Cruciferae in 1928, specifically *Crepis capillaris* (Hollingshead 1928) and *Solanum nigrum* (Jorgensen 1928). Haploids were subsequently found in *Gossypium barbadense*, *Zea mays*, and *Oenothera rubricalyx* (Gates 1929). Anther culture method was created by Guha and Maheshwari in 1964 to produce haploids in a lab setting. In tobacco (Burk et al. 1979) and barley (Kasha and Kao 1970), haploid formation through wide crossing has been documented. For doubling haploid production, the most responsive plant species include barley, rapeseed, and tobacco. More than 250 species have now been studied using doubled haploid techniques.

Parthenogenetic embryos have occasionally been shown to develop from eggs or, less frequently, from synergids (Lacadena 1974). Botanists have also long experimented with different methods to artificially induce parthenogenesis and raise its frequency which is typically only a few plants per thousand by artificial means. The use of foreign and X-rayed pollen, growth regulators (or other substances), delayed or early pollination, and high- or low-temperature shock were some of the treatments used (Kasha 1974). The theory behind the use of foreign or X-rayed pollen was that, although normal fertilization would be inhibited, parthenogenesis might be induced by hormones that diffused from the pollen (Fig. 1.1).

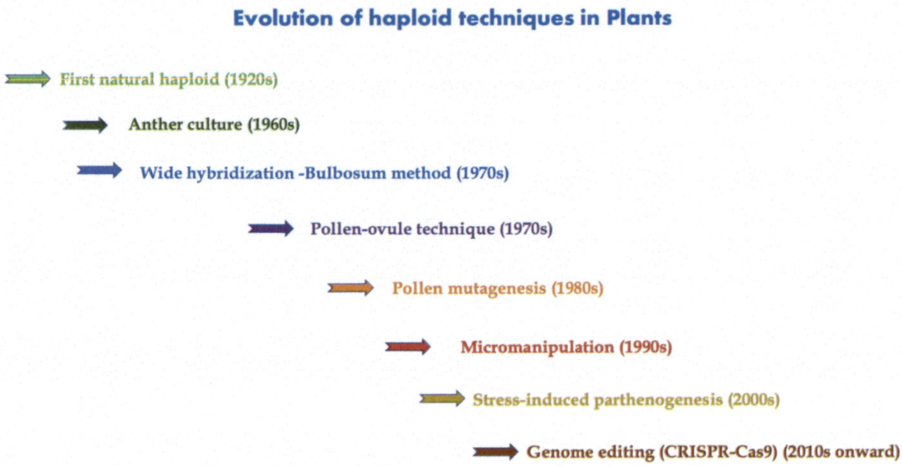

Fig. 1.1 Evolution of haploid techniques in plants

1.2 Methods and Techniques for Haploid Production

Over the last six decades, numerous straightforward and effective methods particularly for in vitro production have been developed to generate vast quantities of haploid plants. Many attempts have been made to induce haploids of higher plants since the first reports of haploid plants in the early 1920s. These efforts can be broadly divided into two categories: in vitro culture which appears to be more effective than in vivo methods and in vivo induction of haploids by various physical, chemical, or biological stimulants. Doubled haploid (DH) technology which depends on the capacity to produce haploid cells that differentiate into haploids or doubled haploid embryos and plants is a significant strategy to speed up genetic gain through a shorter breeding cycle. For a haploid generation, two different approaches can be used: in vitro (culturing of immature male or female gametophytes) and in vivo (development of haploid embryo on a plant) (Ren et al. 2017). Parthenogenesis, in which a seed develops without fertilization, and uniparental genome deletion are the two general methods for in vivo Haploid Induction (HI) (Comai and Tan 2019). Sparse pollination, pollen irradiation, parthenogenesis, seed selection with twin embryos, and wide hybridization are specific in vivo techniques for artificial HI (Kasha and Maluszynski 2003; Touraev et al. 2009). An embryo develops spontaneously from an unfertilized egg cell during parthenogenesis (reviewed in Vijverberg et al. 2019; Bhowmik and Bilichak 2021). The process of apomeiosis (the development of the embryo in the absence of meiosis) occurs during the autonomous or pseudogamous endosperm formation (without or with central cell fertilization, respectively). Apomixis, or clonal seed production, is the general term for this process (Conner et al. 2017). The most well-known example of how apomixis is genetically controlled is the PsASGR-BabyBoom-Like gene from apomictic *Pennisetum*

(Conner et al. 2015). This section mainly discusses progressively in terms of methods and techniques for haploid production.

1.2.1 An Overview of the Methods and Techniques for Haploid Production

In self-pollinated crops, a conventional breeding process using the pedigree technique of selection typically takes 7 years to generate a variety (excluding the testing phase). Numerous drawbacks of traditional breeding are addressed by doubled haploids. It is the quickest path to homozygosity and can be used in breeding programs at any generation. Dominance effects on selection are absent in DH lines since they are homozygous. As a result, DH breeding speeds up breeding initiatives and cuts the time it takes to introduce cultivars by up to 3–5 years. In the case of cross-pollinated species, these DH lines can be employed as parents in breeding programs to generate inbred lines for the generation of hybrid seeds, or they can be released directly as a variety for general cultivation. Additionally, the DH lines can be helpful germplasm for producing new cross combinations.

Expanding DH for F1 hybrid development to high-value crops including aromatic and medicinal plants is currently a topic of great attention. Breeding in these species has been ignored; therefore DH for F1 production should lead to considerable improvements in yields of raw biochemicals that are high, stable, and predictable for the pharmaceutical and nutraceutical sectors (Ferrie 2006). Additionally, DH is a crucial tool for achieving homozygosity in species that are suffering from inbreeding depression, which implies that they will find it difficult to produce fertile homozygous lines through self-pollination. Examples of these species include rye (*Secale cereale*) (Immonen and Anttila 1996), as well as perennial woody species, which are typically characterized by a long juvenile period, a high degree of heterozygosity, large size, and frequently self-incompatibility (Germana 2006, 2009) and generally in plants that have barriers to repeated selfing, such as dioecy and self-incompatibility, where homozygous breeding lines cannot be obtained through traditional methods. In the dioecious crop species *A. officinalis*, sib crossings between pistillate and staminate plants result in 50% male and 50% female offspring, creating an inbred population. However, a homogeneous male population with low-fiber spears is one of the commercially attractive traits of this crop. This species was haploidized using anther culture, and the haploids were then diploidized to create homozygous male offspring. We refer to these as supermales.

1.2.1.1 Introduction of In Vitro Techniques

Androgenesis (Anther or Microspore Culture) The process that produces an embryo or zygote that only has chromosomes from one male parent is known as androgenesis. When androgenesis occurs, callus development either directly or indirectly induces embryogenesis in anthers or microspores. Androgenesis is the process of isolating haploid cells from immature pollen grains by growing the anther or microspores in an artificial medium. These cells are subsequently chromosomal doubled using colchicine to develop DH.

The first people to obtain haploid embryos in vitro by *Datura* anther culture were Guha and Maheshwari in 1966. Dr. J.P. Nitsch began to continue the experiment by recreating the entire haploid plants after seeing the broad possibilities of these remarkable findings. In 1967, he developed haploid *Nicotiana*. He believed that these plants would be very useful for genetic research and plant breeding. Several researchers have now obtained haploid plants from numerous species by culturing anthers; these are divided into two groups: (a) True embryogenetic development is followed by plants that are directly generated from pollen; many species of *Nicotiana*, *Datura*, and *Atropa belladonna* are examples of this. (b) Callus-regenerating plants: A few other plants, including *Brassica*, *Oryza*, *Lycopersicon*, *Petunia*, *Lolium*, *Solanum nigrum*, *Asparagus*, and *Solanum tuberosum*, have also produced positive results in this group.

A detailed examination of DH creation using this technique is outside the purview of this review because there are already several good studies on the topic (Dunwell 2010; Forster et al. 2007; Germana 2011a, b). However, it should be mentioned that haploid creation via in vitro culture is a highly technical process that requires a lot of labor, takes a long time, and is expensive. More importantly, however, it depends on the species and genotype. Over 250 types of microspore embryogenesis have been documented as a consequence of coordinated research on a wide variety of plants (Maluszynski et al. 2003a, b, c; Dunwell 2010). But only a limited range of annual and non-woody plants can successfully use androgenesis.

Male gametophytic cells can be cultured in vitro during the microspore or immature pollen growth stages to produce haploid plants. In vitro, these cells proliferate as haploid calluses or go through embryogenesis in response. Plants can be made again by culturing haploid embryos and inducing shoots and roots in haploid calluses. Through chromosome doubling procedures applied to haploid cells or plants, homozygous, doubled haploid, pure breeding lines can be produced. The first *Datura inoxia* plants were grown by Guha and Maheshwari (1964) using anthers that contained immature pollen. More than 200 species have since been documented to produce haploid plants. In some species, isolated microspores can be cultured to create haploid offspring. In the production of haploid embryogenesis in *Brassica napus* microspore culture, the process has been refined to frequently create high-frequency haploid embryos from this species. There are several useful applications for microspore culture (Dunwell 1985). The biggest benefit is that developing new types takes less time. Using microspore culture can shorten the time it takes to develop a variety by 3–4 years, compared to 10 years for a typical plant breeding program. Second, in the homozygous state, characteristics can be quickly fixed. Since homozygous, doubled haploid lines may be created in a single generation, real breeding lines no longer require 3–4 years of selfing or backcrossing. Third, there is also an improvement in the efficiency of selection. Dominance effects do not disguise the plant's phenotype. Recessive gene-conditioned traits are easily recognized. Fourth, compared to traditional diploid populations, a substantially smaller sample of doubled haploids is needed to screen for suitable recombinants. Moreover, haploid plants or embryos can be doubled for research on biochemistry, physiology, genetic engineering, and mutation. By microinjecting DNA into embryos and using the *Agrobacterium tumefaciens* system, transgenic *B. napus* plants have been created from microspores or proembryos. By selecting microspores in vitro, herbicide-tolerant plants have been created. Haploids are useful for estimating genetic variation and the number of genes for quantitative features, as well as for detecting linkage and gene. Microspore culture is influenced by several factors. These include the culture environment, pre-treatments, media components, donor plant growth circumstances, genotype, and pollen developmental stage. Depending on the genotype and species, conditions vary. The limited rate of embryogenesis and regeneration, the high incidence of albinism, the segregation distortion, and the infrequent doubling of chromosomes to produce DH are additional limitations linked to the application of this technique (Dunwell 2010). The production of DH in barley, *Brassica*, oat, rice, and triticale has been standardized and frequently carried out using this method (Dunwell 2010; Forster et al. 2007; Germana 2011a, b).

Gynogenesis—Ovary/Ovule Culture (Unfertilized Egg Cell) In this process, an embryo develops straight from an unfertilized ovary, embryo sac cell, or ovule. When androgenesis fails, gynogenesis is typically chosen as a backup strategy in species. Consequently, its function in cereals is restricted, despite the fact that it was generated in specific genotypes of rice, barley, and wheat. Because apomictic parthenogenesis and gynogenesis are similar, our knowledge of the mechanisms governing autonomous (i.e., occurring without fertilization) embryogenesis can help advance in vitro gynogenesis techniques. Gynogenesis may be determined by or involve genes that initiate apomictic embryo development from unfertilized egg cells, a process known as parthenogenesis. This may lead to new developments in the field of in vitro gynogenesis. Currently, two cultivated genera *Allium* sp. and *Beta* sp. practice gynogenesis in breeding; however, several species such as barley, gerbera, maize, mulberry, squash, sunflower, and tef are receptive. Onion gynogenesis is accomplished through flower culture on a solid medium.

Recent advancement has been greatly aided by researchers from the Agricultural University of Kraków. A conventional approach was applied to about 100 onion genotypes/accessions, and the results showed a high degree of genetic diversity in the gynogenic response (ranging from zero to ten green plants regenerated per hundred cultured blooms). The majority of the regenerated plants seemed to be haploid, and the DHs needed to be produced by artificial chromosomal doubling. After testing amiprophosmethyl (APM), trifluraline, colchicine, and oryzalin, it was shown that the most effective treatment was a medium containing 50 μM APM. Prior to ovary culture, the induction medium contains polyamines, and the embryos are subcultured prior to being transferred onto the regeneration medium. Although it is challenging to compare different protocols because they haven't been tested on a shared set of genotypes, they have produced valuable breeding material and have an average efficiency suitable for use in breeding programs.

In an early attempt to produce doubled haploid lines in *Beta vulgaris* beets, gynogenesis and androgenesis were both involved. For sugar beet, red beet, and fodder beet, gynogenesis induced from isolated ovules seemed to be successful. The fundamental protocol was created in the latter part of the 1980s. The recent method improvement might be considered to be the cold treatment of inflorescences at 8 °C for 1 week, followed by a reasonably high temperature (30 °C) during the induction phase. In breeding programs, including hybrid breeding, beet-doubled haploids are utilized. It is recently described a comprehensive procedure has been tested in high-throughput DH production. In some trees, sugar beet (*Beta vulgaris*), onions (*Allium cepa*), and other plants, gynogenesis can produce haploid embryos from egg cells in the absence of fertilization (Michalik et al. 2000; Weich and Levall 2003). Furthermore, by coculturing isolated wheat zygotes with embryogenic pollen, a technique for the in vitro regeneration of zygotes and early embryos in wheat may be developed (Kumlehn 2016). Onion and sugar beet gynogenesis require a specific growing temperature for the donor plants, and culture is carried out on induction before regeneration medium is added. Gynogenesis in trees is accomplished by pollinating them with radiation-contaminated pollen from relatives. Although gynogenesis is still the only option for HI for certain plant species, its overall effectiveness is rather poor (Forster et al. 2007; rew. in Bhowmik and Bilichak 2021).

Irradiated Pollen Certain fruits, vegetables, and ornamental plants have been bred to produce maternal haploids by the application of irradiated pollen (Murovec and Bohanec 2012). Although irradiation pollen cannot fertilize an egg cell, it can encourage the formation of the pollen tube within the style and promote germination of pollen (i.e., embryo development) on the stigma. However, because it requires in vitro embryo rescue and is labor-intensive and infrequent, its application has been restricted.

Wide Hybridization In this method, two distantly related (outside of the immediate gene pool) parents are crossed. The chromosomes of the pollinator parent are eliminated due to their nonhomology with those of the female parent, and the resulting embryo contains chromosomes of the female parent only, which are subsequently doubled. In certain Poaceae and Pannicoidae interspecific or intergeneric crossings, paternal chromosome removal from the hybrid embryo occurs after fertilization. These crosses result in poorly developed or non-formed endosperm, which prevents the embryos from maturing during the caryopsis. As a result, in vitro culture and embryo rescue are required. Potatoes are also used for intergeneric crosses, although the process involves parthenogenetic induction of the egg cell rather than chromatin removal.

Wide interspecies crossover is a highly successful haploid induction technique that has been applied to a number of domesticated species. It involves both interspecific and intergeneric pollinations and takes advantage of the haploidy found in the female gametic line. After pollination by distantly related species, plants can produce haploid embryos, as demonstrated by wheat (Laurie et al. 1990) and barley (Kasha and Kao 1970). Haploid production is the result of widespread hybridization in barley between wild *H. bulbosum* ($2n = 2x = 14$) and cultivated barley (*Hordeum vulgare*, $2n = 2x = 14$) as the male. The first haploid induction technique to generate a significant number of haploids spanning the majority of genotypes was called the "bulbosum" approach. Researchers and barley breeders around the world swiftly embraced and refined the technique (Devaux 2003). At lower frequencies, haploid barley plants might also be produced from maize pollen. Barclay (1975) was the first to discover wheat haploids in large crosses between *Hordeum bulbosum* and the wheat variety Chinese Spring. However, because the main crossability inhibitor genes Krl and Kr2 are present in wheat and are expressed in a wide range of wheat varieties, the method was genotype-specific. Subsequently, the wheat × maize system for haploid induction was described (Laurie and Bennett 1987).

The haploid embryos cannot grow normally and the endosperm aborts. In order to keep these haploid embryos at a stage appropriate for isolation and in vitro culture, 2,4-dichlorophenoxyacetic acid is added in situ. With this technique, a huge number of haploid wheat plants of any genotype can be produced. Several other cereals, including barley, triticale (× Triticosecale) (Wedzony 2003), rye (Secale cereale) (Immonen and Tenhola-Roininen 2003), and oats (Avena sativa) (Rines 2003), can also be successfully pollinated with maize to produce haploid embryos. While a number of species, including *H. bulbosum*, sorghum, and pearl millet, have been investigated as pollinators, maize (*Zea mays*) is currently the most often used pollen provider. The wheat × maize haploid production system is highly effective; however, the cost of haploid production is increased since maize must be produced under carefully regulated conditions to coincide with wheat blooming. Thus, for producing the high frequency of haploid and doubled haploids in wheat and triticale, the wheat × *Imperata cylindrica* system was demonstrated to be an effective substitute for the wheat × maize system (Chaudhary et al. 2005; Chaudhary 2007, 2008a, b, 2010). *I. cylindrica* is a winter plant whose flowering season naturally corresponds with that of wheat and triticale. Furthermore, *I. cylindrica*'s genotype is nonspecific for hybridization with any variety of wheat, triticale, or their derivatives, and during the wheat hybridization time, its pollen is easily accessible and abundant.

In potatoes (*Solanum tuberosum*), a comparable mechanism is in place (De Maine 2003). In this instance, the diploid *Solanum phureja* is crossed with the tetraploid ($2n = 4\times$) cultivated potato, *S. tuberosum*. In these crosses, the nuclei of both sperms can fuse with the central cell of ovule to create a functioning endosperm that starts and promotes the unfertilized eggs' parthenogenetic development. Unfertilized egg cells can grow into dihaploid embryos after pollination if they are supported by a 6× endosperm that is created when the polar nuclei of both reduced sperm cells fuse (De Maine 2003). It is possible to increase the diversity of potato genes and more effectively introduce desirable characteristics, including disease resistance, from wild solanaceous species into domesticated potatoes (Rokka 2009).

In-vivo Maternal Haploid Induction Typically, in vivo maternal haploid induction is used to create DH lines in maize (*Zea mays* L.) (Bordes et al. 1997; Prigge et al. 2012). A portion of 8–10% of the growing seeds contains a haploid embryo after pollination of the donor germplasm using pollen inducers (Roeber et al. 2005). Currently, the dominantly inherited marker gene R1-nj carried by the inducer is mostly responsible for distinguishing haploid seeds from diploid F1 seeds produced by normal cross-fertilization. The scutellum and aleurone of diploid seeds become purple due to the marker gene. As a result, it can be utilized as an embryo- and endosperm-specific marker to identify seeds that are haploid versus diploid as well as those that result from accidental outcrossing. Furthermore, the embryo marker's inconsistent color intensity makes it difficult to classify seeds with absolute certainty. If the source germplasm contains the inhibitor alleles such as C1-I (Paz-Ares et al. 1990), which inhibits the expression of the R1-nj marker (Coe 1962), the classification approach is not successful. Another technique based on variations in oil content resulting from pollination with high oil inducers has been devised for quick and accurate differentiation of haploid from diploid seeds (Melchinger et al. 2013). Any maize germplasm can be used with this method, and automated high-throughput screening can be achieved.

Generation of Early Doubled Haploid Wu et al. (2014a, b) observed the appearance of early doubled haploids (EDHs), which they directly produced in maize using in vivo haploid induction. Using flow cytometry and SSR fingerprinting, they further verified the EDHs' homozygous diploid status. At haploid induction during embryonic development, EDH spontaneously doubled. Complete fertility was demonstrated by the EDH lines, suggesting that DH breeding might be utilized as a novel maize breeding approach. However, a number of issues need to be resolved before the EDH system is used in maize breeding practically. These include accelerating the rate at which spontaneous DH is produced, identifying inducers that generate enough EDH, clarifying the mechanisms underlying haploid induction and doubling, or figuring out how spontaneous haploid doubling occurs (Wu et al. 2014a, b) (Fig. 1.2).

1.3 Chromosome Doubling

Spontaneous Chromosome Doubling

In plants generated from anther or microspores, chromosomal duplication can happen spontaneously in cultures frequently in a significant amount removing the requirement for chemically forced doubling. Some people prefer plants that have spontaneously doubled because they are afraid that the process of duplication could result in unwanted alterations. The in vitro settings utilized to stimulate androgenesis may have a significant impact on this percentage. Even yet, we and other writers (Kasha 2005; Kasha et al. 2006) believe that duplication is unlikely to occur spontaneously because a variety of in vitro and ex vitro variables could be at play.

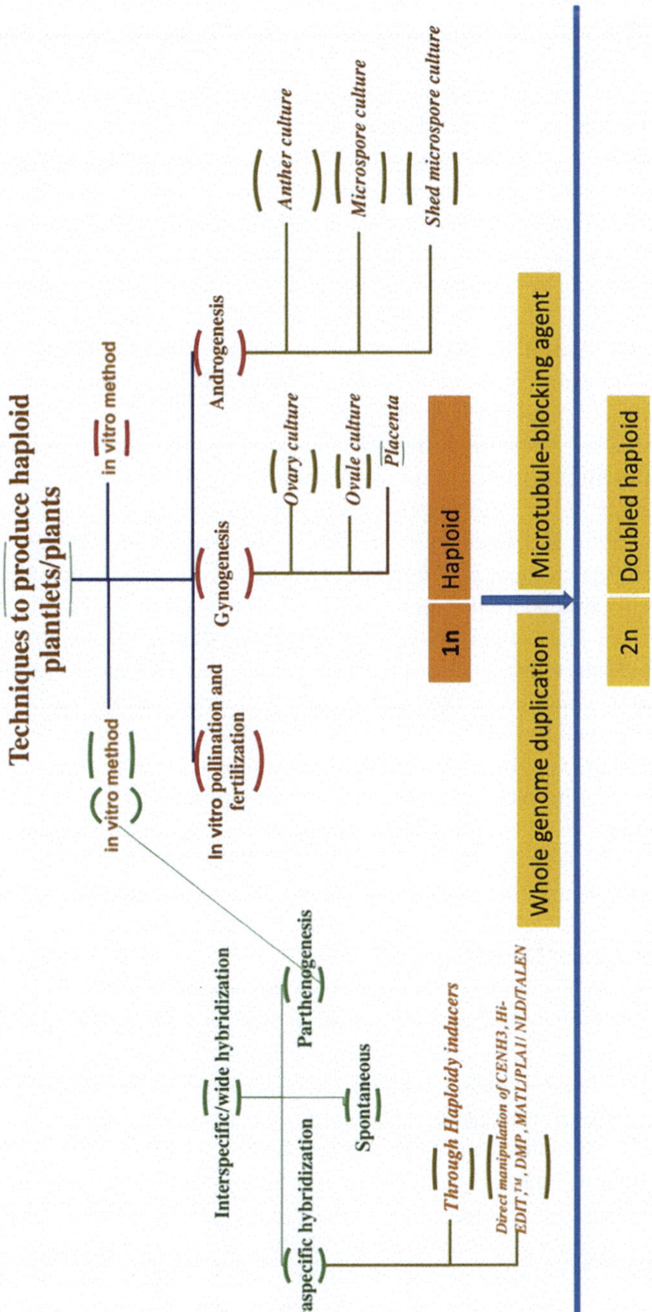

Fig. 1.2 Major techniques to produce haploid plants

The microspore stage at the time of anther/microspore inoculation and the stress treatments or culture conditions used to induce androgenesis, can alter the frequency of chromosomal doubling as noted by Kasha (2005). Numerous factors have been found to affect the frequency of genome duplication, including the length of the inductive conditions, temperature (both hot and cold), mannitol pre-treatments, and colchicine and other antimitotic drugs used to induce androgenesis (Zhao and Simmonds 1995; Henry 1998; Kasha et al. 2001; Zhou et al. 2002a, b; Kasha 2005). DNA duplication events have also been directly linked to the use of plant hormones for in vitro cultures (Joubes and Chevalier 2000; Magyar et al. 2005). An influence can also come from the kind of explant that is employed to induce androgenesis. Comparable outcomes have been observed in *B. oleracea* (Wang et al. 1999) and *B. napus* (Lichter et al. 1988). Endoreduplication has been proposed for a long time as a possible explanation for the spontaneous chromosomal doubling and polyploidy observed in some androgenic animals (Segui-Simarro and Nuez 2008 for specifics on endoreduplication in androgenesis).

Nonetheless, it has been suggested that many of the features of cells normally undergoing endoreduplication are not present in the early phases of androgenesis induction and microspore embryogenesis (Segui-Simarro and Nuez 2008). In the past few years, there has been more evidence supporting nuclear fusion during androgenesis. However, there is unambiguous evidence of nuclear fusion in several species, primarily grains. Pollen-derived plants can sometimes be distinguished from other types of plants using ploidy-level analysis due to the spontaneous doubling of chromosomes that occurs in haploid calli and embryos. Diploid plants grown using the somatic tissue can be either heterozygous diploids or homozygous doubled haploids when in vitro anther or ovary cultures are carried out. For instance, anther culture can be used to produce clonal plant propagation in a variety of genotypes and somatic embryos (Germana 2003, 2005; Germana 2011a, b). Regardless of ploidy, isozyme studies, RAPD markers, sequence-characterized amplified region (SCAR), and simple sequence repeats (SSRs) have been used to evaluate homozygosis and identify the gametic origin of callus and regenerated plantlets.

Induced Chromosome Doubling

Every time an embryo is induced, some of them never double and end up as haploid plants. The number of individuals within a species might vary greatly. There are variations in genotypes within a species as well. A doubling method is necessary for maternally derived haploid plants, as their spontaneous doubling rate is sometimes significantly lower or non-existent. Treatment using anti-microtubule medications, such as colchicine (a natural alkaloid derived from various species of the family *Colchicum*, mainly *C. autumnale*), is the most commonly used application. By binding to tubulin, colchicine suppresses microtubule polymerization. Colchicine is still the most commonly used doubling agent despite being extremely poisonous, requiring a millimolar concentration, and being more effective in animal tissues than in plant tissues. Additional alternatives include trifluralin, pronamide, amiprophosmethyl (APM), and oryzalin, which are herbicides that work well at micromolar concentrations (Rao and Suprasanna 1996). Depending on the genotype, several application techniques are used for successful duplication. According to Caperta et al. (2006), colchicine concentration and therapy time appear to be important factors. For the encouragement of either chromosomal doubling followed by cell cycle advancement or c-metaphase arrest, the proper timing and concentration combination are essential. In terms of doubling frequency and preventing polyploidies or malformations in the embryo, short-term colchicine exposures appear to be more effective than longer treatments (Chen et al. 1994; Zhou et al. 2002a, b). Another element influencing colchicine's ultimate effect is genotype. The most popular times to administer the treatment are (1) to the entire haploid plant or only certain sections after regeneration, (2) to haploid embryos obtained from microspores in the process of growing, and (3) as part of the induction media during androgenesis induction. Haploid plants should only be submerged in colchicine solutions for a few hours at a time. Nevertheless, this technique produces low doubling efficiency and may result in a high incidence of plant deformities and/or mortality. One other disadvantage is that a significant amount of toxic waste is produced, which needs to be disposed of appropriately. Applying colchicine to particular plant parts like roots, axillary buds, or apical meristems is an option. Although these in planta methods are simple, they do not seem to be a very effective long-term approach. Colchicine can also be administered in vitro to calli (Wan et al. 1989) or developing androgenic embryos, although this approach is associated with low survival rates and chimaerism. Applying colchicine directly as early as feasible in the early phases of haploid embryogenic development, however, appears to be more effective. Applying it before to embryo/callus formation might significantly reduce the likelihood of producing chimeric persons.

Furthermore, early application is far more cost-effective because it uses less reagent and produces less waste. Colchicine was applied for 48 or 72 hrs during the first hours of microspore culture in wheat, and colchicine concentrations of 1 mM for 24 h or 0.3 mM applied for 48 or 72 hrs improved the doubling rate (Hansen and Andersen 1998; Soriano et al. 2007). This led to a significant increase in the proportion of doubled haploids in triticale (Wurschum et al. 2012). But when it comes to melons, in vivo colchicine application has been proven to be superior to in vitro treatments in terms of higher doubling efficiency (Yetisir and Sari 2003; Solmaz et al. 2011).

When compared to in vivo and in vitro colchicine as well as in vitro oryzalin inclusion, in vitro trifluralin treatment of *Brassica napus* microspores produced a greater doubling efficiency (Klíma et al. 2008). To produce doubled haploid lines in *Brassica* sp., trifluralin has been proposed as a more effective, less expensive, and less toxic alternative to the commonly used antimitotic drug colchicine (Abraha et al. 2008). Chromosome doubling in a range of species mostly cereals has been induced by N_2O. In dicotyledons, N_2O is thought to be less effective than colchicine.

Chromosome doubling was first observed in tomato (Lindstrom 1929, 1941; Lindstrom and Koos 1931). Colchicine, an antimitotic compound, was isolated in the 1930s and became the standard procedure (Newcomer 1941; Toole and Bamford 1947; Singsit 1991; Lanaud 2006a, b; Lim and Earle 2008, 2009). It is interesting to emphasize that colchicine has also been used to generate haploids in several species, such as sugar beet (Levan 1945), sorghum (Sanders and Franzke 1962; Simantel and Ross 1964), and maize (Hu et al. 1991) (reviewed by Dunwell 2010).

1.4 Advancements in Haploid Production

1.4.1 Introduction of Molecular Markers for Haploid Identification

Marker-assisted selection (MAS) and DH technology both offer the ability to reduce the time required for cultivar development. In contrast to the latter, which may pyramid numerous agronomically beneficial alleles simultaneously into improved genetic background and minimize field testing in early generations, the former only needs two seasons to develop homozygous lines (or one season if microspore engineering and a new breeding technology such as clustered regularly interspaced short palindromic repeats [CRISPR] are employed). When combined with improved molecular breeding techniques, MAS can be utilized as an indirect selection technique to decrease selection costs while simultaneously accelerating and improving the precision of genetic advancement over a shorter number of generations. Therefore, combining MAS with DH opens up new possibilities for creating better selection techniques that maximize selection gains and quicken the development of agricultural cultivars. Additionally, the optimal selection strategy for gene stacking appears to be MAS F_2 enrichment, in which carriers of target alleles are kept in the population and F_2 homozygous individuals for non-target alleles are discarded. This is followed by MAS for high marker score values among DH lines derived from the selected F_2 individuals. Cereal crops have benefited from the effective application of DH and MAS technologies to increase host plant resistance. Resistance to *Fusarium* head blight (*Fusarium graminearum*) can be found, for instance, in wheat stem rust (*Puccinia graminis* f. sp. *tritici*) and stripe rust (*Puccinia striiformis* f. sp. *tritici*) in barley (http://barleyworld.org/sites/default/files/bcd47amba.pdf) and resistance to blast (*Magnaporthe oryzae* B. Couch) and white-backed plant hopper (*Sogatella furcifera*) in rice. One of the most destructive diseases affecting cabbage (*Brassica oleracea* L. var. *capitata*) globally is *Fusarium* head blight (*Fusarium oxysporum f.* sp. *conglutinans*). Lv et al. (2014) successfully developed DH lines combining improved agronomic characteristics and resistance to *Fusarium* head blight by using MAS and microspore culture (androgenesis). These lines, when crossed with elite cabbage inbred lines, produced excellent high-yielding *Fusarium* head blight-resistant hybrids. Using the previously established sources and markers for haploid selection, MAS is also being utilized in maize breeding to create new haploid inducer lines (Kebede 2020).

In maize, the quantitative feature is known as haploid induction rate (HIR). Many closely linked markers were developed by fine-mapping a major quantitative trait locus (qhir1), which affects in vivo haploid induction (Prigge et al. 2012), to a 243 kb region based on analysis of the B73 reference genome using a large F_2 population derived from the cross 1680 × UH400 (Dong et al. 2013). In a more recent study, Dong et al. (2014) developed potential high oil inducer lines with an oil content of around 8.5% and an HIR of about 8% using these markers. This suggests that MAS for qhir1 together with HIR selection was useful in speeding up haploid induction in maize. These days, the Centro International de Mejoramiento de Maíz y Trigo (CIMMYT, Mexico) combines host plant resistance to the maize streak virus with adaptability through DH-based MAS (Prasanna et al. 2012).

1.4.2 Genomic and Transcriptomic Studies on Haploid Plants

Plant breeding programs can become more effective using genomic selection. This method estimates genomic estimated breeding values (GEBVs) for performance prediction by combining phenotyping with genome-wide marker data. Genomic selection takes into account all marker information, as opposed to MAS, which only takes into account strong marker-trait relationships. This helps capture more of the variation caused by small-effect QTL and prevents biased marker effect estimations. Two different kinds of data sets are used in genomic selection: training and validation sets. To find prospective germplasm or cultivars with expected genomic value (predicted based on phenotype and GEBVs of the training populations), a training population with known GEBVs is required to advance breeding programs in the future. Lines that have undergone both genotyping and phenotyping make up the training population, whereas lines that have undergone genotyping alone are used to predict GEBVs. Numerous factors influence the prediction of the genomic selection accuracy, such as trait heritability, linkage disequilibrium, training population size, marker type and quantity, the relationship between training and test populations, and genotype × environment interaction. In maize, simulation studies have unmistakably demonstrated that the response to DHL-based genomic selection was larger than that of F_2 populations or marker-assisted recurrent selection (MARS) with different numbers of QTL (Mayor and Bernardo 2009). There are more sources with methods and statistical methodologies for calculating GEBVs.

The testcross performance of 1380 DHLs of maize from 36 crosses that were previously genotyped with 1152 SNPs and involved in multi-location trials for grain yield and grain dry matter content, was used for checking the accuracy of genome-based prediction. Based on genomic data, they found that the average prediction accuracy for grain yield was good (0.72–0.74) among related families, but it drastically dropped (0.47–0.480) when predictions were made among distantly related families. Additionally, prediction accuracy dropped with decreasing sample size, and it remained high when population size was cut in half (0.67–0.69).

The prediction accuracy for lines from individual crosses is influenced by the makeup of training populations. In maize, 635 DHLs from five interconnected crossings with four parents were genotyped using 16,741 SNPs, the agronomic attributes of these DHLs. The genetic similarity pattern across the five populations clearly distinguished is between full siblings, half-siblings, and unrelated groups, reflecting the crossover method. When taking into consideration the impact of trait heritability and sample size, the prediction accuracy within full-sib families nearly matched theoretical expectations. However, when half-sib DHLs were substituted for full-sib DHLs, the accuracy fell by 42%. If half-sib DHLs from both parents in the validation population were available, rather than just one, statistically substantial improvements in the results might be obtained. When utilized alone or in conjunction with related families, the unrelated crosses that displayed opposing linkage phases with the validation population produced lower or negative prediction accuracies, respectively. The genomic prediction results in maize breeding at CIMMYT in parallel. Thus, it is evident that DHLs are appropriate for genomic selection.

Due to their perfect homozygosity and lack of heterozygosity, DH lines are the most valuable and significant material for quantitative inheritance studies. Their suitability for researching "quantitative inheritance" is thus optimal (Touraev et al. 2009; Mochida et al. 2004). DH lines can be used to study key aspects of quantitative inheritance, such as the number of genes regulating a quantitative trait, gene interactions between distinct genes, gene linkages, and chromosomal placements (Choo 1981a; Jui et al. 1997). By calculating the square of divergence of extreme DH plants from the sample mean and dividing it by genotypic variance or additive variance of DHs, one can calculate the number of genes governing a polygenic trait (Choo 1981b). To investigate gene linkage, DH lines from the F_1 and F_2 generations are created. Then, the means, variances, and skewness of the two types of DH lines are compared to provide an understanding of the gene linkage. Snape and Simpson introduced this approach in 1981.

1.4.3 Haploid-Based Trait Mapping and Genetic Studies

For basic research in fields including genomics, gene expression, and genetic mapping, DHs are frequently utilized (Ferrie and Caswell 2011). Because of their complete homozygosity, ability to repeat the experiment multiple times at different labs, and shorter time necessary to establish a big DH population, DH populations are perfect for genetic mapping. Because of a straightforward segregation ratio (1:1) akin to backcross, creating maps with a DH population resulting from an F_1 of two homozygous parents is quite simple (Snape 1976). Furthermore, scoring the polymorphism under complete homozygous conditions in multiple contexts over some time is possible with the use of DH lines for QTL mapping (Young 1994).

In many species, including barley (*H. vulgare*), rice (*Oryza sativa*), rapeseed (*B. napus*), and wheat (*T. aestivum*), DHs have proven essential to the establishment of genetic maps (Forster and Thomas 2005a, b). Genetic analysis in out-pollinating species can be made simpler by employing at least one DH parent in the initial cross to create a segregating population. This method has worked well for vegetables like Brussels sprouts, cauliflower, cabbage, and broccoli in the *Brassica oleracea* complex. Plant breeders are particularly interested in trait mapping, which is made possible via marker maps. Quick deployment in QTL research resulted from the creation of molecular marker maps in DH populations. Within a year following the release of their corresponding marker maps, the first barley QTL investigations were published (Heun 1992; Hayes et al. 1993). MQTL software has been created for QTL analysis at several places for complete homozygous lines (e.g., DH lines) (Tinker and Mather 1995). The procedure of creating a segregating DH population from an F1 hybrid to identify marker-trait relationships has become the standard practice in sensitive crop species, such as barley (Tuvesson et al. 2006). QTLs for biotic stress, yield, quality, agronomic traits, and developmental traits have all been identified in rice using DH population mapping (Forster and Thomas 2005a, b). Similar to this, QTLs have also been found in pepper, rapeseed, flax, wheat, barley, maize, mustard, and so on (Forster and Thomas 2005a, b).

In bulked segregant analysis (BSA), which compares individuals from two extremes of population distribution for a given trait, such as disease resistance and susceptibility, DHs are also perfect for establishing marker-trait relationships. Similar to quantitative trait mapping, robust phenotyping is necessary for accuracy in BSA. Since DHs may be tested repeatedly, trustworthy data, including field data from trials spanning multiple seasons, can be collected. A variety of marker systems can then be used to gradually genotype the DNA from the chosen DHs. Marker-assisted selection for many breeding traits, primarily disease and pest resistance, but also quality attributes, has been successfully established by BSA and DH analysis (Michelmore et al. 1991).

DHs are significant in the field of genomics. To find potential genes, expressed sequence tags, or ESTs, are employed. With the aid of BAC libraries, it is possible to determine the precise chromosomal location of these ESTs as well as map their position about QTLs. To precisely target candidate genes, DH populations are essential for the integration of genomic and physical mapping. In the IR 64× Azucena DH map of rice, ESTs are co-located with important genes and QTLs for resistance to major illnesses (Wang et al. 2001). A further technique for connecting genes to phenotypes is induced mutation. For both forward and reverse genetics, TILLING (Targeting Induced Local Lesions in Genomes) mutant populations have been created in several species. To prevent false positives due to inherent diversity in the starting material, mutant populations must be obtained from an inbred or homozygous line, preferably a DH line.

1.4.4 Production of Transgenic Plants

Transgenes can be integrated into the haploid microspore system to create transgenic plants that are homozygous. Desired genes can be transformed using both haploid and diploid DNAs, leading to the production of stable transgenic plants that can be utilized for genetic research such as gene mapping and identification. (Forster et al. 2007). While unicellular microspores are the main target, cells or explants at all phases of microspore embryogenesis and regeneration have been employed as gene delivery receivers; the selection process is based on the potential for regeneration following treatment.

Numerous transformation methods, such as particle bombardment, electroporation, microinjection, and *Agrobacterium tumefaciens*-mediated transformation, have been tested (Touraev et al. 2001). The first homozygous transgenic plants were

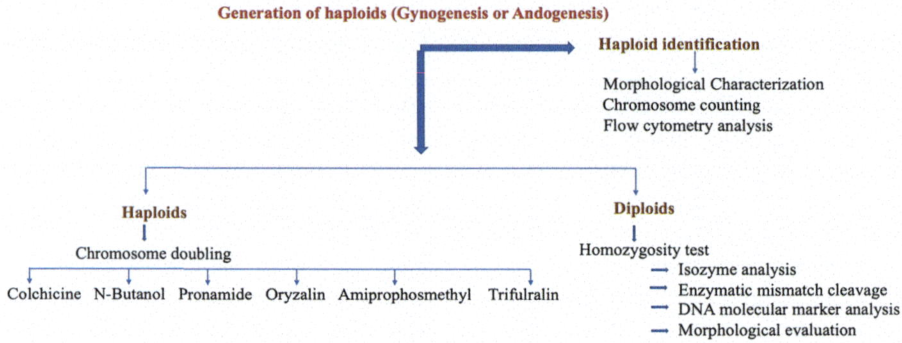

Fig. 1.3 Emerging techniques and technologies for haploid production and analysis

created by Jahne et al. (1994) by biolistic blasting of barley microspores. Particle bombardment of tobacco has targeted immature pollen and both embryogenic and non-embryogenic microspores (Stoger et al. 1995; Touraev et al. 1997a, b). Through the use of the transgene (HVAI), DH transgenic wheat plants were created with increased resistance to simulated water stress. Over several generations, the transgene was expressed and steadily incorporated (Chauhan and Khurana 2011). The issue with the transformation method is that transgenic plants must be stabilized for homozygosity once a transgene is introduced to examine the phenotypic impact of the gene transfer. More time and money are needed for this. Therefore, DH in conjunction with transgenic technologies can improve transgenic efficiency and accelerate transgenic production (Fig. 1.3).

1.5 Applications of Haploid in Crop Improvement

The study and use of haploid cells continue to be important in various fields including genetics, agriculture, and biotechnology as they provide insights into the mechanisms that generate genetic diversity. The release of the rapeseed (*Brassica napus*) cultivar Maris Haplona (Thompson 1972) and the barley (*Hordeum vulgare*) cultivar Mingo was obtained in Canada in the late 1970s. This was the first crop plant released using DH technology. The development of nearly 300 new varieties has resulted from the routine use of DH technology by numerous breeding companies and laboratories worldwide that are involved in the improvement of a wide range of crops (Tuvesson et al. 2007). These data are likely underestimated due to the lack of information from some private companies and some parts of the world (Thomas et al. 2003; Dunwell 2010). Barley (96 variants) has the highest number of DH varieties, followed by rapeseed (47), wheat (20), and the remaining types in rice, pepper, tobacco, eggplant, melon, triticale, and asparagus. Nowadays, DH cultivars are found in many crop species. For example, barley (*H. vulgare*) is thought to yield 50% of its modern cultivars in Europe (Forster et al. 2007). DHs are often employed as parents for the development of F1 hybrid seeds in vegetable crops.

The current status of haploids is very promising. Scientists have made a significant progress in developing new and improved methods for producing and maintaining haploid cells. This has led to an increased interest in using haploids for a variety of applications including:

Use of Haploids in Plant Breeding and Crop Improvement
Haploids can be used to develop new crop varieties more quickly than traditional breeding methods. This is because haploids are homozygous meaning that they have two identical copies of each chromosome. This makes it easy to identify and select desirable traits. Techniques like anther culture and microspore culture have been refined and applied to create homozygous plants faster.

Backcross Breeding Backcross conversion, which is typically transferred from a hardy or wild genotype (donor) to a high-yielding but susceptible genotype (receiver) of the same species, is a classic plant breeding technique for improving an elite line lacking in a particular characteristic like disease resistance. To reconstitute/restore the genes of high yield in the new cultivar, this is accomplished by backcrossing the F1 hybrid to the recipient parent repeatedly during the first year of crossing the donor and recipient parents. To recover the recipient parent's DNA, 5–6 years are needed. Finding a gene of interest or disease resistance at each generation is another issue with backcross breeding, especially if the disease resistance is regulated by a recessive gene.

In this instance, to produce recessive homozygotes for gene expression, we must self each BC generation. Compared to the dominant gene, this requires twice as much time. In contrast to the issues with the traditional backcross approach, the use of DH technology in conjunction with molecular markers not only addresses the aforementioned issues but also considerably shortens the time required for product development. In a backcross generation, a molecular marker is utilized to identify a gene of interest, and DH is created to achieve homozygosity in a single generation. In barley (*H. vulgare*), the technique has been effectively applied for introgressive stripe rust resistance (Chen et al. 1994; Toojinda et al. 1998). Backcrossing results in backcross inbred lines (BILs), and DH from this backcross is highly beneficial and perfect for gene mapping. Therefore, the combination of double haploidy and marker-assisted selection (MAS) improves the accuracy and efficiency of selecting desired features. Additionally, resistance gene stacking benefits from it.

Mutant Selection One of the most significant uses of haploid development methods has been the induction and selection of mutants using haploid systems (Kasha 1974). In seed-propagated crops, the isolation of a homozygous mutant line typically takes three to four generations; however, the DH system can cut this time in half. Following mutagenic treatment, all altered traits manifest at both haploid and doubled haploid levels in the first generation. The phenotypic-based DH mutant lines are genetically fixed and do not segregate in the following generation. Because of the homozygous condition for mutants, the frequency of the desired mutant in the DH population will be significantly higher than in the standard M2 generation.

Haploids appear to be a useful platform for in vitro mutant selection. By screening a large mutagenized population in vitro, the haploid cell population can offer a chance to find a rare mutation event. Maluszynski et al. (1996) have proposed two methods for introducing mutagenesis into haploid cells in vitro. The first scenario involves directly applying a mutagenic treatment to haploid cells or tissues; the alternate method involves treating seeds with a mutagen and using M1 plant gametes as donors for the haploid culture. Microspores present a compelling substitute for treating mutations. It is possible to directly generate a population of homozygous mutant lines by directing the mutation treatment at individual gametic cells, inducing embryogenesis, and DH plant production (Szarejko and Forster 2007). This allows for the screening of both recessive and dominant mutants in the first generation. Because of the gametic segregation ratio of 1:1 and the absence of chimerism, DH also improves the efficiency of selection for desired mutants. The drawback is that homozygous fatal mutations that would otherwise remain in the heterozygous state would be eliminated.

Reverse Breeding and Male Sterility DH plants are essential to reverse breeding, a relatively new technology. Wijnker et al. (2007) proposed the methodology for "reverse breeding." With this approach, DHs are recovered from plant microspores when little to no recombination has taken place. First, among the population that is segregating, the superior hybrid genotypes are found. Next, a gene for the induced suppression of meiotic recombination is added via genetic transformation. Molecular markers can be used to screen the resulting recombinant inbred populations and identify those with complementary chromosomal combinations. This will enable the reconstruction of an initial heterozygous parent of the DH through hybridization of the two people. Consequently, it is possible to rebuild the existing F1 hybrids by identifying distinct parents with distinct chromosomal constitutions.

Tobacco has been utilized to produce F1 hybrids through reversible male sterility and doubled haploid generation. Based on the metabolic engineering of glutamine in growing tobacco anthers and pollen, the technology was created. Tobacco was employed to inactivate cytosolic glutamine synthetase (GS1) by introducing mutant tobacco GS genes coupled to the NTM19 promoter unique to microspores and the TA29 promoter specific to tapetum. Male sterility resulted from pollen in primary transformants aborting near the first pollen mitosis. By using microspore embryogenesis, a non-segregating population of homozygous doubled haploid male-sterile plants was produced. Plants were sprayed with glutamine to restore fertility, or they were pollinated with pollen that had grown in vitro in a glutamine-containing media. An inventive, ecologically friendly breeding technique is produced when twofold haploid production is combined with reversible male sterility.

Haploid Maize Doubled by "Inducer" Line

Inbred "inducer" lines were used in a technique that produced very promising results in maize (Sarkar et al. 1994; Shatskaya et al. 1994; Chalyk 1999). In recent times, Röber et al. (2005) have created and characterized a new and improved inducer line as "RWS." Approximately 10% of the progeny that use it seem to be haploids descended from the mother plant. Additionally, RWS has a gene for a color marker that allows kernels with haploid embryos to be distinguished from those with diploid embryos. Numerous loci are implicated in the ability to produce haploid embryos, according to QTL analysis, albeit the precise genes involved are unknown. Microscopy revealed that following RWS self-pollination, some embryos' nuclei lose their chromatin. Three-week-old embryos removed from growing caryopses display a high number of micro-nuclei in shoot primordia (Wedzony 2003). Breeders are becoming more and more interested in the technique as practical methods for haploid induction and chromosomal duplication have been devised (Deimling and Fleihinghaus-Roux 1996; Röber et al. 2005).

The bulbosum approach, also known as the in vivo method (interspecific hybridization) for haploid induction, was initially applied to barley (Kasha and Kao 1970). The fundamental process that results in the production of a haploid embryo in the early stages of a hybrid embryo is uniparental chromosome elimination. It has also been documented that several species in the genus *Hordeum* can produce haploid offspring when crossed with *H. vulgare* (Houben et al. 2011; Jorgensen and von Bothmer 1988). In tobacco (Burk et al. 1979) and potatoes (Caligari et al. 1988; Peloquin et al. 1996), interspecific hybridization has also been used to isolate DH. *Avena sativa* × *Z. mays* (Kynast et al. 2012), *Triticum turgidum* × *Z. mays*, wheat × pearl millet (*Pennisetum glaucum* (L.) R.Br) (Laurie 1989), *Triticum aestivum* × Triticeae species (*H. vulgare* and *H. bulbosum*), and *T. aestivum* crossed with more distantly related species (Liu et al. 2014) have all been used to isolate DH. For the generation of DH lines in wheat, hybridizations between wheat and barley or maize and wheat are recommended (Jauhar et al. 2009). The hybridization between wheat and barley has a high frequency (76%) of wheat maternal haploids and is a useful resource for studying the mechanism of uniparental chromosome elimination in cereals.

The most popular method for producing haploids in maize is in vivo maternal haploid induction. Maternal haploids genomes come only from their seed parent, and pollen is the source of haploid induction. Due in major part to the discovery of high inducer stocks (WS14, MHI, CAUHOI, and RWS), which when crossed with other lines, release up to 10% haploids, this system has significantly improved over time in maize (Wu et al. 2014a, b, and references therein). Inducer lines like PHI# 1, 2, 3, and 4 have been used in maize recently, and the results have produced 11–16% haploids, albeit the reaction varies depending on the genotype.

The most often used chemical to double chromosomes is colchicine, which is administered to seedlings in vivo. It is an anti-microtubule medication that binds to tubulin to prevent microtubule polymerization. Its high cost, toxicity, and labor-intensive nature sometimes pose significant obstacles to the large-scale manufacturing of DH lines, despite its efficacy in artificial chromosomal doubling. According to Häntzschel and Weber (2010), anti-microtubule herbicides such as pronamide, oryzalin, and amiprophosmethyl can effectively double the number of chromosomal sets in maize while being less harmful than colchicine. It has also been discovered that nitrous oxide gas effectively doubles the number of chromosomes in maize (Kato 2002). While rare, spontaneous DH incidence has been shown in maize (Geiger et al. 2006). This should be further studied to find lines that naturally produce DHs at high frequency to avoid chromosome doubling due to colchicine. An alternative to employing microspore culture to double haploid production is the in vitro administration of colchicine. Wurschum et al. (2012) demonstrated a noteworthy rise in the percentage of haploid triticale plants that were doubled using this technique of applying colchicine, so offering a potentially beneficial substitute for the in vivo method.

Wu et al. (2014b) observed the appearance of early doubled haploids (EDHs), which they directly produced in maize using in vivo haploid induction. Using flow cytometry and SSR fingerprinting, they further verified the EDH's homozygous diploid status. At haploid induction during embryonic development, EDH spontaneously doubled. Complete fertility was demonstrated by the EDH lines, suggesting that DH breeding might be utilized as a novel maize breeding approach. However, a number of issues need to be resolved before the EDH system is used in maize breeding practically. These include accelerating the rate at which spontaneous DH is produced, identifying inducers that generate enough EDH, clarifying the mechanisms underlying haploid induction and doubling, or figuring out how spontaneous haploid doubling occurs (Wu et al. 2014b). Some fruits, vegetables, and ornamental plants have used irradiated pollen to produce maternal haploids (Murovec and Bohanec 2012). Although irradiation pollen cannot fertilize an egg cell, it can encourage the formation of the pollen tube within the style and promote germination of pollen (i.e., embryo development) on the stigma. However, because it requires in vitro embryo rescue and is labor-intensive and infrequent, its application has been restricted. Zhang et al. (2007) synthesized DH (SynDH) in wheat by utilizing meiotic restoration genes. It is necessary to cross an alien species without the need for embryo rescue and an inducer line with the gene(s) for meiotic restitution. The process involves three steps: hybridization to trigger recombination, interspecific hybridization to separate haploids, and spontaneous chromosome doubling through selfing the interspecific F1. No specialized tools or therapies are needed to produce DH.

Potato Double Haploidy Through S. phureja × S. tuberosum Crosses

By pollinating with the diploid potato species *Solanum phureja*, tetraploid genotypes of *Solanum tuberosum* (cultivated potatoes) can yield doubled haploids (Mendiburu et al. 1974; De Maine 2003). The two male sperm cells of *S. phureja* participate in the creation of functional endosperm in around 0.5% of pollinated ovules. This process initiates the parthenogenetic development of unfertilized egg cells. Hybrid *S. tuberosum* × *S. phureja* seeds may be easily separated from seeds carrying haploid embryos since the best pollinator lines of *S. phureja* were developed for a dominant purple spot embryo marker. Recent developments in chromosome number duplication techniques have improved the efficiency of potato production through androgenic methods (Jacobsen and Ramanna 1994; Rokka et al. 1995; Tai and Xiong 2003). Additionally, compared to crosses with *S. phureja*, androgenesis is applicable to a much wider range of *Solanum* species (Jacobsen and Ramanna 1994; Rokka et al. 1995; Aziz et al. 1999).

IG1 System for Paternal Haploid Induction in Maize

The use of haploid induction technology in maize breeding dates back many years. Two main strategies are used for in vivo haploid induction: the first uses the indeterminate gametophyte1 (ig1) mutant, while the second relies on Stock6-derived lines. Parthenogenesis, or the ig1 mutation, has also been reported in plants treated with hormones or following genetic manipulations. Haploidy can also occur after pollination with (defective) pollen treated by radiation, chemicals, or high temperatures. In certain in situ methods, embryo rescue is necessary because endosperm development is not occurring.

First observed as a spontaneous mutation in the inbred line Wisconsin-23 (W23), haploid embryos or seedlings undergo spontaneous genome duplication or are treated with a microtubule-blocking agent such as colchicine to induce the reduplication of the haploid genome and the generation of doubled haploid plants. This results in approximately 3% haploid induction rates (HIR) of paternal haploids. As a female parent, the inducer containing ig1 is employed. Thus, the haploid genome of the pollen donor and the cytoplasm of the female inducer are present in paternal haploids. Maternal haploids were also produced using ig1 mutants (Kermicle 1969), though their frequency (about 0.1%), which depended on the genetic background, was too low to be distinguished from the haploids that spontaneously formed from crossings of wild-type plants (Chase 1949).

In the process of breeding maize, the capacity to create paternal haploids has been used to move chromosomes from one variety of the crop into the cytoplasm of another variety (Ren et al. 2017). Ig1 is located on maize chromosome 3 and encodes a protein with the LATERAL ORGAN BOUNDARIES (LOB) domain, according to fine-mapping studies (Evans 2007). LOB domain-containing proteins are members of a broad family of transcription factors peculiar to plants that have been demonstrated to be necessary for the development of lateral organs in higher plants (Husbands et al. 2007). The closest *Arabidopsis* homolog of IG1, ASYMMETRIC LEAVES 2 (AS2), is crucial for maintaining adequate primary and secondary vein development as well as proper leaf left-right symmetry. A Hopscotch retrotransposon insertion has been found in the second exon of the maize gene, upstream of the encoded LOB domain, in ig1 mutant plants of the inbred line W23 (Evans 2007). Several phenotypes were seen throughout the development of the female gametophyte (embryo sac), despite the fact that ig1 mutants are viable. Due to aberrant microtubular behavior and errors in mitotic synchronization, decellularized ig1 embryo sacs have an excess of nuclei. Individual cells in the micropylar portion of the embryo sac, which typically consists of one egg and two synergid cells, continue to go through mitosis after the embryo sac has cellularized. Moreover, aberrant phragmoplasts, asynchronous microtubular patterns in various nucleus pairings, and irregular nuclei positioning have all been noted (Huang and Sheridan 1996).

An extension of the proliferative phase results in the production of more egg cells, central cells, and polar nuclei inside central cells. About 75% of embryo sacs that are perforable by two or more pollen tubes have three or more synergid cells. There have been reports of multiple sperm and egg cell fusions occurring within the same embryo sac. This may involve the creation of several embryos. According to the traits listed above, wild-type IG1 helps the maize embryo sac transition from proliferation to differentiation (Evans 2007; Guo et al. 2004; Huang and Sheridan 1996). The discovery that IG1 is a target of the egg cell-secreted differentiation factor EAL1 signaling pathway lends credence to this theory (Krohn et al. 2012). Many damaged ig1 embryo sacs result in aberrant kernels containing two or more embryos, tiny endosperms, or early seed abortion because of their atypical structure (Evans 2007; Kermicle 1969).

Unknown precise pathways underlie haploid induction resulting from ig1 mutations. It is well known that proteins with LOB domains interact with transcription factors belonging to the basic helix-loop-helix (bHLH) family and control the expression of numerous downstream genes. It has been demonstrated that in leaf primordia, the *Arabidopsis* homolog AS2 suppresses the expression of the knotted1-like homeobox (knox) genes AtKNAT1, AtKNAT2, and AtKNAT6 (Semiarti et al. 2001). It has been demonstrated that the expression of genes such as OsEG1, OsMADS6, and OsMADS1 in the embryo sac is impacted by the downregulation of the rice homolog OsIG1. This suggests that OsIG1 may be involved in controlling genes linked to the development of floral organs and female gametophytes (Zhang et al. 2015). It has also been revealed that maize ig1 embryo sacs downregulate KNOX8, suggesting the existence of common gene regulatory mechanisms (Evans 2007). The precise mechanisms, however, are yet unknown, which makes it challenging to apply the IG1 system to other crops.

Stock6-Derived Maternal Haploid Induction Systems of Maize

Stock6-derived haploidy inducer lines are the second above-mentioned method that is widely used for in vivo haploidy induction in maize. According to Coe (1959), Stock6 is an inbred line that initially caused haploid induction rates of 2.3–2.2%. Several research teams and breeders went on to incorporate this founder line's haploid-inducing ability into diverse genetic contexts. Additional hybrids were created, such as a cross between Stock6 and the aforementioned ig mutant in W23, which served as the foundation for several inducer lines that are currently in use, including RWS, UH400, MHI, and PHI, which had haploid induction rates of between 7% and 16% (Prigge et al. 2012).

A mapping strategy was utilized to identify a QTL related to haploid induction at chromosome 1 bin 1.04 using progeny from a hybrid between the inducer line PK6 (haploid induction rate 6%) and the non-inducer line DH99. It was designated as ggi1 or gynogenesis inducer 1. A decreased rate of transmission via male gametes is likely the cause of incomplete penetration and segregation distortion that were found (Barret et al. 2008). However, a more recent study found that the haploid induction ability of lines produced from Stock6 is a complicated quantitative feature regulated by several loci. Four biparental populations were examined: three non-inducer lines, the temperate line 1680, the two tropical lines CML395 and CML495, and an inbred UH400 that served as a pollinator against another haploid-inducing line, CAUHOI. As a result, eight QTLs on six chromosomes were discovered. Prigge et al. (2012) reported the detection of a mandatory QTL for haploid induction on chromosome 1 to bin 1.04, which was dubbed qtl for hir 1 (qhir1). This finding is consistent with prior investigations. On chromosome 9 (qhir8) is located another significant QTL. This clarifies the 20% observed genotypic variance and supports the notion that additional loci that re unable to induce haploid induction on their control qhir1 (Liu et al. 2015; Prigge et al. 2012).

Through detailed mapping using the progeny test approach, the 50.3 Mbp-large qhir1 locus was further reduced to 243 kbp. In these studies, line 1680 was pollinated by inducer line UH400. The progeny underwent two rounds of selfing before being pollinated to produce hybrid ZD958, which allowed haploids to be identified using the R1-nj marker that was produced from UH400. According to Dong et al. (2013), the novel QTL known as qhir1 is required for haploid induction. A genome-wide association analysis (GWAS) was carried out utilizing 1482 non-inducers and 53 haploidy induction lines in an additional attempt to pinpoint the sub-region within qhir1 that is important for haploid induction. qhir12 is a QTL of 3.97 Mbp common for all inducers, which was absent in all non-inducers and which did not overlap with qhir11. Another group employed selfed progeny of recombinants of a large F2 population obtained by crossing the tropical haploidy inducer line TAIL8 with the tropical non-inducer CML269 to dissect the sub-region of qhir1. The qhir11 sub-region, out of the two qhir1 sub-regions, significantly influenced the production of mother haploids without the need for the qhir12 allele, while the "inducer" qhir12 allele did not result in haploid induction at a rate greater than the spontaneous rate seen in wild-type cobs. Furthermore, it was noted that the qhir11 area exhibited kernel abortion and segregation distortion, both of which are traits of haploidy inducer lines (Nair et al. 2017).

More recently, three different groups independently reported that haploid induction in Stock6-derived lines requires a patatin-like phospholipase found in qhir11. MATRILINEAL (MTL), PLA1 (PHOS-PHOLIPASE A1), and NOT LIKE DAD (NLD) were the names given to the phospholipases (Gilles et al. 2017; Kelliher et al. 2017; Liu et al. 2017). The MTL/PLA1/NLD gene's fourth exon contains a 4-bp insertion that was found in inducer lines that are characterized by a protein truncation at amino acid 380. This location is followed by 20 modified amino acid residues and an early stop codon that truncates the protein by 29 amino acids as a result of the insertion. It has been demonstrated that MTL/PLA1/NLD mutants can cause haploids in vivo, along with the usual side effects of segregation distortion and kernel abortion (Gilles et al. 2017; Kelliher et al. 2017; Liu et al. 2017). For instance, CRISPR-Cas9 lines that targeted the phospholipase's first exon exhibited 9–14% kernel abortion and an average of 2% haploid induction (Liu et al. 2017). In Stock6-derived lines produced by transcription activator-like effector nuclease (TALEN)-mediated targeted mutagenesis, small deletions close to the region of the

4-bp insertion resulted in haploid induction of 4–12% (average 6.65%) (Kelliher et al. 2017).

MTL/PLA1/NLD activity in maize is limited to ripe pollen grains and pollen tubes, according to gene expression research (Gilles et al. 2017). Sperm cells in germinated pollen tubes localize to the cytoplasm and plasma membrane (PM) of wild-type MTL/PLA1/NLD, which has a C-terminally fused fluorophore. Interestingly, regardless of whether an endogenous or constitutive promoter was selected for its expression, the shortened form does not express at detectable levels (Gilles et al. 2017; Kelliher et al. 2017). This finding implies that the mutant form in maize cells is unstable. The enzymes that catalyze the hydrolysis of acyl groups from phospholipids to create free fatty acids and lysophospholipids are members of the phospholipase A (PLA) superfamily. MTL/PLA1/NLD is one of these enzymes. Depending on the hydrolysis site's specificity, plant PLAs are classified into three families: PLA1s, PLA2s, and patatin-like phospholipases (pPLAs/PLPs) (Chen et al. 2013; Scherer et al. 2010). Plant pPLA/PLP family (Gilles et al. 2017; Kelliher et al. 2017) includes MTL/PLA1/NLD. Based on sequence similarities, this family is further subdivided into three sub-classes. Class I phospholipases have an unidentified function in the N-terminal domain and a leucine-rich repeat domain in the C-terminal, which exhibits the strongest homology to animal iPLAs. Class II proteins are more closely linked to patatins than class III proteins, which have a distinct catalytic center with the motif G×G×G as part of the catalytic dyad (Chen et al. 2013; Scherer et al. 2010). MTL/PLA1/NLD is a member of class II pPLAs/PLPs, according to phylogenetic study (Gilles et al. 2017). We still don't know the exact methods by which MTL/PLA1/NLD's C-terminal truncation induces haploidy. Furthermore, the impact of the truncation on MTL/PLA1/NLD function remains unclear. According to others, the loss of membrane binding capacity might be crucial (Gilles et al. 2017). Wild-type MTL/PLA1/NLD-CITRINE localized to the cytosol and plasma membrane (PM) when expressed in the *Arabidopsis* root epidermis under the constitutive AtUBQ10 promoter, with signals assigned to tiny intracellular compartments. The shortened form only showed signals in the cytosol and lost its PM localization. The shortened protein's loss of putative S-palmitoylation or S-farnesylation sites at C423 (Gilles et al. 2017) lends even more credence to this theory.

A potential phosphorylation site may also be eliminated as a result of the C-terminal protein truncation, which could have a significant impact on the protein's capacity to be activated during cellular signaling. Phosphorylation of the C-terminus of two of the closest *Arabidopsis* homologs of MTL/PLA1/NLD, AtPLAIVA/PLP1, and AtPLAIVB/PLP5, by calcium-dependent protein kinases (CDPKs), has been demonstrated to impact their function.

AtCDPK3 totally eliminated the phosphorylation of AtPLAIVA/PLP1 and AtPLAIVB/PLP5 by deleting the final 16 amino acids, which contained four serines at the C-terminus. A mutation of Ser399 (equivalent to Ser402), conserved in MTL/PLA1/NLD, significantly reduced the efficacy of protein phosphorylation. It has also been demonstrated that AtPLAIVA/PLP1 and AtPLAIVB/PLP5 phosphorylation considerably impacts their activity and substrate selectivity (Rietz et al. 2010).

Therefore, it's likely that MTL/PLA1/NLD also has to be phosphorylated at its C-terminus by an unidentified kinase specific to sperm cells in order for cell signaling responses during fertilization to properly transduce signals.

In either case, haploid induction and kernel abortion result from modulating MTL/PLA1/NLD function, regardless of whether this modulation affects the protein location, activity, or substrate specificity. There have been suggestions that these phenomena are (i) related to either endosperm identity in the central cell (resulting in defective endosperm followed by kernel abortion) or failure of fertilization while zygote identity is activated in the egg cell (haploid induction via parthenogenesis) or (ii) related to post-zygotic uniparental elimination of the paternal genome. In conclusion, the exact mechanism by which Stock6-derived lines and mtl/pla1/nld mutants pollinate haploid induction remains a mystery. Studies on single fertilization (Tian et al. 2017) as well as uniparental chromosome removal (Li et al. 2009, 2017a, b; Zhang et al. 2018; Zhao et al. 2013) are currently in progress. How qhir8, the other major QTL identified in maize haploidy inducers (Liu et al. 2015; Prigge et al. 2012), influences MTL/PLA1/NLD function is an intriguing subject as well. Progeny from the cross of two inducers with a fixed qhir1 region, CAUHOI (2% haploid induction) and UH400 (8% haploid induction) were analyzed to fine-map qhir8.

It was discovered that F2 plants homozygous for UH400 at qhir8 had a substantially greater haploid induction than F2 plants homozygous for CAUHOI at qhir8. According to reports, the heterozygous F2 plants' mean induction rate fell between that of the two homozygous classes. According to Liu et al. (2015), these findings demonstrated that qhir8 has the capacity to quadruple the haploid induction rate resulting from qhir1. Subsequent investigations will clarify the intricate molecular processes behind MTL/PLA1/NLD activity in both mutant and wild-type plants.

It will be interesting to see if site-directed alteration of orthologous genes in other crops can induce haploid induction in the mother, which would allow the strategy to be used in crops without functional haploid induction systems. Using RNA-guided Cas endonuclease technology, a pollen-specific MTL/PLA1/NLD homolog (OspPLAIIφ or OsMATL) underwent site-directed alteration as a preliminary step toward haploid induction in rice. According to Yao et al. (2018), knockout mutants displayed a decreased seed set and a haploid induction of up to 6%. Given how encouraging this initial finding is, it seems likely that haploid induction will soon be achievable in additional grass species as well. Significant future efforts are required to boost haploid induction and spread this crucial breeding characteristic to eudicotyledonous crops. While obtaining adequate haploidy induction rates requires the use of effective inducer lines, it is also important to consider the significance of the "donor" genotype. Chase noted in 1949 that the rates of spontaneous haploidy induction in maize are significantly influenced by the genotypes of both parent plants (Chase 1949). Furthermore, it has occasionally been observed that a preponderance of variables imparting low frequency of haploidy exists (Coe 1959; Lashermes and Beckert 1988).

Eder and Chalyk (2002) examined the impact of source germplasm on the haploidy induction rates of two paternal inducer lines, MHI and M741H, generated from Stock6. Four European flint lines, 11 dent lines, and 5 flint × dent hybrids were among the examined maternal genotypes. All genotypes were capable of inducing maternal haploids, albeit at varying rates, and no genetic pool had a clear advantage over the others. The source germplasm is crucial for tropical maize as well, since it works with the environment to produce variations in HIR for the same paternal inducer that range from 2.9% to 9.7% (Kebede et al. 2011).

Two loci on chromosomes 1 and 3, designated qtl for maternal hir 1 (qmhir1) and qmhir2, respectively, are thought to exist and account for up to 20% of the phenotypic variation, according to mapping efforts aimed at determining the maternal contribution to haploidy induction (Wu et al. 2014a, b). A similar, if less obvious, relationship was found for indeterminate gametophyte 1 between the genotype of the pollinator and the frequency of maternal and paternal haploids (Lashermes and Beckert 1988).

Generation of Haploids via Induction of Parthenogenesis: BBM-Triggered Maternal Haploid Induction

As previously said, it is still unknown if the MTL/PLA1/NLD system causes parthenogenesis to be induced or results in the deletion of one or both uniparental chromosomes. Parthenogenesis, the process of creating an embryo from an unfertilized egg cell, is another essential element of apomixis, along with apomeiosis, which results in unreduced gametes and autonomous endosperm development in some species (Ronceret and Vielle-Calzada 2015). Numerous plant species have independently evolved apomixis, or asexual reproduction through seeds, but the genes underlying apomixis have remained mysterious.

An important gene linked to the parthenogenesis trait was recently identified in *Pennisetum squamulatum*, an apomictic grass that grows pearl millet. In sexual pearl millet, parthenogenesis may be successfully induced at a rate of approximately 35% by using the gene PsASGR-BABY BOOM-like (PsASGR-BBML), which is expressed in egg cells prior to fertilization (Conner et al. 2015). Homologous genes are not present in the egg cells of species like maize and rice, but they are promptly activated de novo in the zygotes following fertilization (Anderson et al. 2017; Chen et al. 2017). That the PsASGR-BBML gene may also be utilized to induce parthenogenesis in rice and maize, resulting in the development of haploid embryos at a rate of 25–89%, is therefore not surprising (Conner et al. 2017).

To create haploid embryos, the native pearl millet PsASGR-BBML promoter and an *Arabidopsis* egg cell-specific promoter were both utilized. It was observed that PsASGR-BBML is not adequate to trigger parthenogenesis in the *Arabidopsis thaliana* model plant. Conversely, a similar finding was made: BABY BOOM (BBM), a transcription factor belonging to the AP2/ERF family, was initially identified in *Brassica napus* (oilseed rape) and later demonstrated to be able to induce somatic embryos on *Brassica* and *Arabidopsis* seedlings following ectopic overexpression (Boutilier et al. 2002). It was proposed that BBM is a factor involved in the morphogenesis and proliferation of embryonic cells. Using an embryo sac-specific promoter derived from ES1–ES4 genes, the first attempts to transfer BBM from *Brassica* to maize (Cordts et al. 2001) failed because parthenogenetic embryo development was not induced (*S. Amien* and *T. Dresselhaus*, unpublished results). In early embryogenesis, eudicots and monocots have distinct genetic programs (Zhao et al. 2017). Furthermore, it's possible that the individual transcription factor genes in the BBML family have distinct targets and are therefore not interchangeable. Given that zygotic BBML genes from maize can induce parthenogenesis once expressed from an egg and/or embryo sac-specific promoter, it is intriguing to investigate whether zygotic BBML genes exist in eudicots and whether they can be utilized for haploid induction in *Brassica* and other eudicotyledonous crops.

Generation of Haploids via Interspecific Hybridization
Two distinct parental genomes are fused together into a single nucleus during interspecific fertilization, which is often embedded in the maternal cytoplasm. A genomic and epigenetic rearrangement of the genomes happens as a result of intergenomic parental conflicts brought on by such a unique genomic constitution (Riddle and Birchler 2003). Uniparental genome removal has been reported in more than 100 different species combinations, including 75 and 26 examples for mono- and eudicot species, respectively, despite the fact that most of the time the parental genomes remain integrated at least partially following successful fertilization (Ishii et al. 2016).

For instance, haploid wheat can be produced by pollinating wheat with the pollen of any of the following: *Sorghum bicolor*, *Imperata cylindrical*, *H. vulgare*, *Z. mays*, *Coix lachryma-jobi*, *Teosinte*, *Pennisetum glaucum*, or *Tripsacum dactyloides* (Ishii et al. 2016). Due to endosperm abortion, the majority of hybrids between monocot species needed in vitro tissue culture to save developing embryos. For *Brassica napus*, a very high frequency of doubled haploids has just been discovered. According to research, up to 98% of the haploids produced by pollinating allohexaploid *B. napus* (AAAACC) with pollen from artificially produced allooctaploid *Brassica* (AAAACCCC) doubled.

Uniparental chromosome elimination can occur as quickly as during the first zygotic division or as slowly as 3 weeks after fertilization, depending on the species combination. This was shown in studies involving the combinations *wheat* × *Imperata cylindrica* (Mukai et al. 2015) and *wheat* × *Pennisetum glaucum* (Gernand et al. 2005). The development of strong protocols, such as embryo rescue techniques that can produce haploid embryos in up to 30% of pollinated florets, encouraged the use of the so-called "bulbosum" method (reviewed in Kumlehn and Stein 2014). In recent times, the discovery of the centromeric histone H3 variant CENH3 in *H. vulgare* × *H. bulbosum* embryos revealed that uniparental centromere inactivation plays a role in the process of uniparental chromosome elimination (Sanei et al. 2011). The CENH3-positive active centromeres of *H. vulgare* and *H. bulbosum* were differentiated from the CENH3-negative inactive centromeres of *H. bulbosum*, or the quantity of CENH3 was decreased in unstable hybrids. It is most likely that sperm-derived centromere-incorporated CENH3 proteins sustain *H. bulbosum* residual kinetochore function until it decreases to a level necessary for proper chromosome segregation, ultimately leading to chromosome elimination. With each cell division, the amount of pre-existing CENH3 will be about half if it is divided evenly across duplicated sister centromeres and no de novo incorporation of CENH3 into *H. bulbosum* centromeres occurs. Effective kinetochore assembly can be facilitated in humans by as little as 10% of the endogenous CENH3 (Liu et al. 2006). *A. thaliana* (Ingouff et al. 2010) showed that grasses also exhibit zygotic resetting of CENH3, which would result in the active elimination of both parental CENH3s before the first zygote division and lessen the reactivation of *H. bulbosum* centromeres in unstable hybrids. In unstable barley × *H. bulbosum* hybrids, centromere inactivation is not caused by uniform CENH3 inactivation (Sanei et al. 2011). It is thought that CENH3's integration into the paternal genome's centromeres is compromised in unstable hybrids. Numerous proteins regulate CENH3 loading and assembly into centromeres; if any of these proteins operate incorrectly, it could lead to a non-functional centromere (reviewed in Gohard et al. 2014). But aside from KNL2 (Lermontova et al. 2013) and GIPs (Batzenschlager et al. 2015) from *A. thaliana*, no other plant protein implicated in the formation and upkeep of CENH3-specific chromatin has been found as of yet.

A defective release of sister chromatid cohesion is probably linked to the centromere-dependent process of elimination in Triticeae (such as *T. aestivum*, *T. spelta*, *T. durum*, *Triticale*, *T. monococcum*, *Secale cereale*, and *Avena sativa*) × pearl mill combinations (Ishii et al. 2010). It is unclear, therefore, if centromere inactivity and the compromised release of sister chromatid cohesion work in tandem.

Crosses with Maize and Related Species

Plant breeders became interested when they read the initial studies (Zenkteler and Nitzsche 1984) about embryo development following wheat-maize pollination. Subsequent research indicated that during the early stages of the hybrid embryo development, maize chromatin is removed from the hybrid nucleus (Laurie and Bennett 1986, 1989). When Pooideae members cross with maize, endosperm is rarely generated, and in plants that have growth regulators, embryos can abort as soon as 5 or 6 days after fertilization if appropriate care is not given (Laurie and Bennett 1989; Brazauskas et al. 2004). Different treatments that preserve embryo growth to the point where they can be isolated and cultured for further growth have been created using synthetic auxin equivalents to spikes (Riera-Lizarazu et al. 1992; Matzk and Mahn 1994; Mujeeb-Kazi and Riera-Lizarazu 1996; Wędzony et al. 1998a, b). Crossability barriers between maize and several cereals were not found, so green plants were successfully obtained in a wide range of genotypes, including those that were resistant to androgenesis, such as barley (Furusho et al. 1991), triticale (Wędzony et al. 1998a, b), rye (Deimling and Fleihinghaus-Roux 1996), and wheat (Almouslem et al. 1998; Inagaki et al. 1998; Saidi et al. 1998; David et al. 1999; Cherkaoui et al. 2000; Knox et al. 2000). Results for barley and triticale are comparable to those of wheat; however in the rye, few embryos are found and are challenging to save. Conversely, oats were simple to cross with maize (Rines 2003). While partially fertile and sometimes retaining maize chromosomes, oat polyploids derived from oat × maize crossings are not good for breeding, but they did yield valuable unique aneuploid stocks for genetic and gene introgression research (Rines 2003). In subtropical and tropical regions, it is possible to match the flowering periods of maize and most Pooideae in the field; but experiments are conducted under glasshouse conditions in Europe and Canada. Sadly, there is no reliable way to preserve the vitality of maize pollen in storage.

Some organisms are related to maize that could serve as pollinators in place of maize. Teosinte (*Zea mays* ssp. *mexicana*), a wild Mexican corn variety, was used in crosses with wheat by Ushiyama et al. (1991). Hexaploid and tetraploid wheat were pollinated using sorghum (*Sorghum bicolor*) (Riera-Lizarazu et al. 1992). Furthermore, some trials successfully used pearl millet (*Pennisetum glaucum*, synonym *Pennisetum americanum*) (Inagaki and Mujeeb-Kazi 1995; Ohkawa et al. 1992). These species are known as "haploid-inducing pollinators," and their use may be advantageous in hot settings where mother plants' and pollen donors' blossoming seasons can readily coincide. Comparative studies typically demonstrate that hybrid maize of the sweet corn kind is a good pollinator.

Agrobacterium-Mediated Genetic Conversion to Androgenesis

Kumlehn et al. (2006) developed primary transgenic (T0) plants in barley carrying a single copy of the sequence integrated using a novel genetic transformation method based on infecting androgenic pollen cultures with *Agrobacterium tumefaciens*. Of these plants, about 60% set seed, indicating spontaneous genome doubling, presents the possibility of producing doubled haploid T1 seeds that are instantly homozygous for the transgene. More importantly, since many angiosperms are capable of androgenesis and *Agrobacterium*-mediated transformation, this technique has a great deal of promise for use in other plant species.

It is possible to generate DH homozygous for the transgene using haploid embryos. Employing explants from anther culture-derived haploid embryos of the commercial wheat cultivar CPAN1676, a drought-tolerant and stable transgenic haploid wheat plant employing the barley gene HVA1 was developed using *Agrobacterium*-mediated genetic transformation (Chauhan and Khurana 2011). Colchicine was used to double the number of chromosomes in transgenic plants, resulting in DHs that did not exhibit transgene silencing until the T4 generation. In addition, these DH plants outperformed non-transgenic DH plants in drought tolerance and exhibited quicker seed germination and seedling establishment. DH technology can also help repair transgenes and eliminate undesirable selectable marker genes at the same time. This method, which conserves time and resources, could be helpful for functional genomics research as well as for creating stable true breeding transgenic-derived DHs with the desired genes.

Transgenic plants were created using unicellular tobacco microspores (Touraev et al. 1997a, b). The male-germ line transformation (MAGELITR) protocol's creators claimed that it is quick, regeneration-independent, and free of somaclonal variation and chimerism. Using a double-cassette vector (RS 128/Xyl) carrying 1,4-β-xylanase or an endochitinase gene obtained from *Trichoderma harzianum*, androgenic microspores of wheat were generated either by electroporation or by co-cultivation with *Agrobacterium tumefaciens* (Brew-Appiah et al. 2013). Eudes et al. (2014) reported that microspore transformation was made easier by the use of short peptide nanocarriers as a delivery system, providing new opportunities for plant breeding. For instance, haploid-transformed triticale microspore plants were regenerated following the genetic engineering of microspores through the coupling of plasmid DNA with cell-penetrating peptides (Chugh et al. 2009). One benefit of this DNA delivery into microspores is that a selectable marker gene is not necessary; instead, an end-point PCR-based technique is adequate for quickly identifying a transformed plant produced from microspores. Targeted genome editing is a common use for transcription activator-like effector nucleases (TALENs), which produce site-directed genetic change in a variety of cells and creatures (Chen and Gao 2013). Gurushidze et al. (2014) used *Agrobacterium*-mediated transformation to express GFP-specific TALEN pairs in the embryonic pollen of transgenic barley carrying a functional copy of GFP to show that customizable designer nucleases are expressed and function in haploid cells, which can then be rapidly regenerated into non-chimeric homozygous mutants. The ability to create DH lines with the desired trait(s) in one season as opposed to the two seasons required for crossbreeding or for elucidating gene function is being made possible by microspore engineering and more recent breeding methods like TALENs.

Haploid Production Using Centromere-Mediated Genetic Engineering
The regrowth of plants from haploid (gametic) tissues, the eradication of a single parents genome (e.g., by radiation of pollen), and the selective loss of one-parent chromosomes during early embryogenesis (Sanei et al. 2011) (which can be induced by interspecific hybridization or by "haploid inducing" variants) are the three ways that haploid plants can be induced. Certain species or cultivars respond better to any of the aforementioned methods than others. None of these strategies are effective for many species. In more recent times, *Arabidopsis* transgenic technology was created to induce uniparental inheritance. This technology has effects similar to those caused by haploid inducers or interspecific hybridization. The transgenic parent's chromosomes are lost selectively during the early stages of embryogenesis. This transgenic technology involves the manipulation of the protein CenH3, a variant of histone 3 that is specifically localized to centromeres, replacing H3 in centromeric nucleosomes. It has been determined that the position of CenH3 defines the position of the centromere, as most organisms define the centromere epigenetically (Allshire and Karpen 2008). The process of experimentally induced re-localization of CenH3 to alternate chromosomal loci results in the formation of functional centromeres at the new site, and, conversely, the formation of new functional centromeres ("neo-centromeres") is accompanied by the migration of CenH3 localization to the site of the new centromere (Burrack and Berman 2012). One crucial stage in chromosomal biology is the selective reloading of CenH3 to centromeric locations following DNA replication. Ectopic loading would produce a dicentric, which can cause a chromosome break during division, but maintaining CenH3 status would result in a chromosome that is unable to segregate. While researching CenH3 function, Ravi and Chan (2010) found a haploid-inducing variation of CenH3. The sequence of centromeric H3 (also known as CenpA), in contrast, varies greatly even between closely related species, whereas non-centromeric histones are generally conserved (Cooper and Henikoff 2004).

There has been much conjecture about this adaptive evolution, which is shared by numerous genes involved in reproduction (Swanson and Vacquier 2002; Henikoff et al. 2001). Similar to other histones, the structure of CenH3 may be separated into two parts: the N-terminal tail, which exhibits the majority of the extreme variability seen even among CenHs from angiosperms, and the histone fold domain, which interacts with DNA and other histones (Cooper and Henikoff 2004). Loosely organized, the N-terminal tail extends into the cellular region around the nucleosome and is frequently tagged with methyl, phosphate, or acetyl groups that impact chromatin activity. Currently, *A. thaliana* is the only model plant for which CENH3-mediated haploid induction techniques that produce a sufficient number of haploids have been devised (Karimi-Ashtiyani et al. 2015; Kuppu et al. 2015; Maheshwari et al. 2015; Ravi and Chan 2010). Except maize, where a rate of up to 3.6% haploids was attained, there is currently no effective CENH3-based haploid technology available for crop species (Kelliher et al. 2016). This goal is currently the subject of intense research, and two methods for CENH3-based uniparental genome deletion are explored. The first method is based on the two-step procedure described by Ravi and Chan (2010), who successfully restored lethal *A. thaliana* knockout mutants with modified CENH3 transgenes through transformation. Later, the same group showed that untagged natural CENH3 variations from more distant relatives, such as various *Brassica* species, can likewise be used to functionally complement *A. thaliana* cenh3 null mutants and induce haploid growth (Maheshwari et al. 2015). The earlier theory that evolutionarily near CENH3s could target centromeres in alien but closely related species was also supported by this investigation (Nagaki et al. 2010). The two-step approach would probably result in transgenic haploidy inducer lines and the need for genetic alterations.

The loss of chromosomes of the transgenic inducer plant is expected to render the resulting haploid plants non-transgenic, but using plant material that has undergone a genetic transformation in breeding programs raises some concerns, particularly in Europe. The alternative one-step method relies on genetic changes of the endogenous CENH3 gene that are relatively focused. Point mutations that were not fatal but caused decreased CENH3 loading to centromeres and/or the formation of so-called weak centromeres in barley, sugar beetroot, and *A. thaliana* were chosen for a series of experiments using TILLING approaches (Karimi-Ashtiyani et al. 2015; Kuppu et al. 2015). In the laboratories of the researchers, targeted mutagenesis employing RNA-guided Cas9 endonucleases is being utilized to modify CENH3 genes to produce partially functional CENH3 alleles, which represents a feasible substitute for the TILLING technique. The possibility of implementing a particularly promising one-step modification of CENH3 without any transgene integration is made possible by the recent establishment of using preassembled complexes of purified Cas9 protein and target gene-specific guide RNAs (Woo et al. 2015), in addition to conventional genetic transformation using RNA-guided Cas endonuclease expression units.

Functional CENH3 genes of the corresponding species must be cloned and examined for the identification of appropriate modification sites within the hypervariable N-terminal region or the more conserved histone fold domain. This is a requirement for haploid formation using inducer lines carrying novel CENH3 variants. Furthermore, if a two-step approach is chosen, the stable expression of foreign or genetically modified CENH3 genes is required to partially compensate for indigenous CENH3 knockout mutations. Thus far, cloning and characterizing CENH3 genes have been done on several species, including rice (Hirsch et al. 2009), tobacco (Nagaki et al. 2009), soybean (Tek et al. 2010), *Brassica oleracea* (Wang et al. 2011), onion and garlic (Nagaki et al. 2012), pea (Neumann et al. 2012), common bean (Iwata et al. 2013), carrot (Dunemann et al. 2014), wheat (Yuan et al. 2015), and rye (Evtushenko et al. 2017). To the best of our knowledge, however, techniques for haploid induction in crop plants other than maize, where mutagenesis was used to induce alterations within the CENH3-coding area, have not yet been published in the peer-reviewed literature. However, haploidy inducers in rice and tomatoes are described in a patent from KEYGENE N.V. (Wageningen, Netherlands) (patent WO 2017/200386 (Op Den Camp et al. 2017)). At amino acid position 9 of the N-terminal region of CENH3, lysine was exchanged for glutamine in tomato due to a non-synonymous mutation of A to G. As many as 4% of progeny showed evidence of being doubled haploids when this haploidy inducer line was utilized as a pollinator of parents that were wild-type. The single nucleotide polymorphism analysis was used to determine the homozygosity of these plants.

Three haploidy inducer lines in rice produced up to 1% of haploid progeny by substituting different single amino acids. Furthermore, according to the Dutch breeding business Rijk Zwaan (patent WO 2017/081011, Van Dan et al. 2017), haploid progeny has been produced in cucumber (*Cucumis sativus* L.) and melon (*Cucumis melo* L.). When haploidy inducer plants with a heterozygous CENH3 frame-shift mutation crossed with wild-type plants, 1% of the progeny produced were haploid or twice haploid. Melon plants with homozygous mutations in the HFD of CENH3 were crossed with wild-type plants at a frequency of 1.5% to produce haploid or doubled haploid progeny. The number of successful applications of CENH3-based haploidization in crop species is opposed to the high number of reviews on the technique (Britt and Kuppu 2016; Comai 2014; Copenhaver and Preuss 2010; Ishii et al. 2016; Ren et al. 2017; Watts et al. 2016). As a result, there are great hopes for this strategy's applicability in crops; however developing workable CENH3-based haploidization techniques is difficult.

Advances in CRISPR/Cas9-Based Double Haploid Induction

CRISPR/Cas systems are novel plant breeding strategies that enable effective target-specific gene editing in crops, in contrast to genetically modified (GM) technologies. According to El-Mounadi et al. (2020), this is expected to accelerate crop development, boost trait-specific yield, and revolutionize agricultural biology in the future. In contrast to ZFN and TALENs, the CRISPR/Cas9 system is the most versatile, effective, and straightforward genome editing technology available. Breeders will be able to create haploid inducer lines that can be used to create parthenogenetic embryos by using the CRISPR/Cas9 system to specifically knock out the CENH3 and DMP genes in crops, even if the application of this technique will face many challenges. But to create haploid inducer lines that have been edited using

CRISPR/Cas9, a successful transformation method would be necessary for at least one genotype. One potential way to improve the efficiency of double haploid generation in resistant species is by the use of haploid inducers in genome editing. Adopting MATL or CENH3 mutant lines for haploid formation and generation in dicots or cereals may promote the obtention of these traits in many species and prevent gynogenesis, androgenesis, or parthenogenesis caused by pollen irradiation (Hooghvorst and Nogués 2020).

By targeting the genes involved in natural fertilization in gynogametic cells with CRISPR/Cas9, fertilization was blocked, and haploid embryos were produced from egg cells (Hooghvorst and Nogués 2020). The haploid CRISPR-Cas9 systems eschewed the need for significant protocol adaptation to many genotypes. Currently, genes like MATL, DMP, and CENH3 that are involved in chromosomal segregation are the focus of haploid inducer lines. The mutant gene that typically produces the haploid inducer ability in cereals is the MATRILINEAL (MATL) gene, sometimes referred to as NOT LIKE DAD (NLD) or PHOSPHOLIPASE A1 (PLA1). It encodes a phospholipase unique to pollen (Prigge et al. 2012). For example, a 4-bp insertion was found in the carboxyl terminus of the MATL gene in natural haploid inducer lines of maize.

In cereal species, the MATL gene is present and highly conserved, but not in dicots. Consequently, the centromeric histone 3 (CENH3) gene is typically the target of the haploid inducer-mediated genome editing machinery in dicots. In cv. Ailsa Craig, a CRISPR-Cas9 mutagenesis construct including the FAST-Red marker for haploid identification was converted to generate slDMP mutants. It has been proposed that the slDMP mutation promotes the formation of haploid in vivo embryos in tomato. This method would be a major improvement over in vitro haploid induction, in addition to additional members of the Solanaceae, Cucurbitaceae, and Fabaceae that are resistant to DH production (Hooghvorst and Nogués 2020). CRISPR-mediated CENH3 has the potential to revolutionize the creation of haploid inducer lines in cucurbit species due to its specificity in numerous sequences of the CENH3 gene that can be easily targeted. Zhong et al. (2022) focused on DMP genes to create maternal haploid inducer lines for the formation of haploid cells in allopolyploid (allotetraploid) *Brassica napus* and *Nicotiana tabacum*.

1.6 How to Identify the Haploidy?

Nowadays, many crops are bred using the DH approach to create commercial hybrids. There are numerous ways to differentiate diploid and haploid plants in higher plants. The most noticeable difference is that they appear smaller, in part due to their smaller cells; generally speaking, the ploidy level of a plant is strongly correlated with its cell volume. Numerous techniques can be employed to verify whether a plant is haploid. These include indirect approaches based on guard cell and plastid dimensions, as well as direct methods of measuring the number of chromosomes using traditional cytological techniques and flow cytometry (Bohanec 2003); field identification; morphological, radioactive, herbicide, molecular, and genetic marker methods; and conventional cytogenetics (Chen and Song 2003). Using the "Navajo" kernel trait, for instance, allows for the identification of haploids by selecting kernels that have a pigmented endosperm and a non-pigmented embryo. However, donor genotype and environment can influence anthocyanin gene expression, which in turn affects color. In the meantime, the operator visual tiredness will cause the screening accuracy to decline.

The availability of molecular markers facilitates the use of genomic selection (GS) and marker-assisted selection (MAS) to quickly screen the breeding population and choose the optimal allele combinations (Ren et al. 2017). For example, trait introgression into an elite variety requires less time and resources when DH and MAS are coupled (Forster et al. 2007). The F1 plants are backcrossed multiple times to the parental line after an elite variety crosses with the line bearing the desired allele. This allows the donor genome to be cleared by recombination and segregation while retaining the target region and applying the DH method as soon as after the first backcross can significantly speed up this laborious and time-consuming approach. The effectiveness of the DH technique for producing viable plants has a major influence on the stage at which DH is applied during backcrosses. Furthermore, when several QTLs influence the trait and the population size is limited, the benefit of employing DH versus F2 populations in marker-assisted recurrent selection (MARS) and GS was higher in maize (rew. by Bhowmik and Bilichak 2021).

Fully Automatic High-Throughput NMR System The NMR approach, on the other hand, is based on the oil content, which is directly measured by NMR (Wang et al. 2016). It can be scaled up to a high-throughput system to meet the demands of a high-volume, practical selection process (Melchinger et al. 2013, 2014; Liu et al. 2012; Dong et al. 2014). For example, millions of seeds can be continuously screened over the course of 2 weeks at a rate of roughly one kernel per second. Non-destructive oil analysis of materials and seeds has been done using NMR (Nordon et al. 2001). Kotyk et al. (2005) analyzed the oil content of 2592 maize kernels arrayed on a cubic sample holder at a rate of one kernel per second using a 1.5-T whole-body MRI scanner with a 1-m diameter bore (Kotyk et al. 2005). A superconducting magnet with a 2.1-T magnetic field and a 30-cm bore can analyze more than 20,000 seeds every hour, or 5 kernels per second, using the continuous wave-free precession (CWFP) sequence (Colnago et al. 2007; Wang et al. 2016).

1.7 Perspective

The study of haploids continues to evolve as new technologies emerge. The advent of CRISPR-Cas9 gene editing has further enhanced the utility of haploids for studying gene function and identifying therapeutic targets. As research in this area progresses, we can anticipate even more significant contributions from haploids to genetics, genomics, and biotechnology. Haploids and diploid hybrids (DHs) are widely used in crop breeding 90 years after the discovery of naturally occurring sporophytic haploids in Jimson weed and roughly 50 years after the breakthrough in producing haploids from anther culture in *Datura* and the use of bulbosum method for haploid induction in barley. The field of DH research has made significant strides and has enabled the production of numerous cultivars, primarily for grains and *Brassica*.

Researchers have gained a great deal of knowledge about the genetics and mechanisms underlying haploid induction, as well as the factors that influence it. They have also discovered markers (morphological, biochemical, and DNA markers) that can be used to identify putative haploids and increase genetic gains in plant breeding by utilizing DH technology. Now, numerous research teams worldwide are evaluating its significance in the field of plant breeding. The main benefit of this technology is that the DH plants that are produced may not be subject to regulations in some markets since their chromosomes are exclusively derived from their non-transgenic parent. By combining DH technology with applied genomics, off-season nurseries, controlled environment facilities, and breeding operations, new opportunities are created to optimize genetic gains in selection, reduce the time required for cultivar development, and increase breeding efficiency. The application of DH technology in the large-scale production of inbred lines for the development of hybrid cultivars has been widely adopted by multinational seed firms. The fact that these facilities while currently scarce have further supported public sector breeding initiatives and small- to medium-sized private seed businesses globally is heartening. These platforms charge for their services, which include complete or partial assistance in haploid induction and application in breeding.

The maize system, initially harnessed by breeders before comprehending the underlying haploid induction mechanism, contrasts with the serendipitous discovery of the CENH3 system through fundamental scientific inquiry, subsequently adapted for breeding purposes. Consequently, ongoing research deciphering the molecular and cellular intricacies of these systems promises to revolutionize fundamental scientific understanding encompassing areas like pollen development, double fertilization, gamete interactions, genome stability, and practical applications in breeding methods. Envisioning the future, one plausible trajectory involves mimicking these mechanisms by manipulating other molecular components. For instance, disrupting centromere proteins beyond CENH3 might destabilize the centromere, theoretically leading to haploid induction. Additionally, exploring alternative methods to induce sperm cell chromosome fragmentation, distinct from MATL/NLD/ZmPLA1 mutations, could potentially yield higher haploid induction rates.

Despite all these advancements, there are still certain issues with DH technology use that need to be resolved. The need to establish DH technology where it is not currently available (for those crops), increased efficiency in DH formation (including automated procedures), and the development of novel strategies based on haploid and/or DHs with regard to breeding of various crop species, germplasm enhancement, management of genetic resources, and the creation of novel types of experimental populations are some of these (Lubberstedt et al. 2015). The genotype-dependent response to androgenesis and the occurrence of albinos in in vitro culture are other pertinent issues that require further research. These include genetic switching from the gametophytic to the sporophytic phase during embryo production without fertilization, in vitro spontaneous production of diploid hybrids (DHs) to avoid the use of hazardous chemicals like colchicine and epigenetic control of gene expression in gametogenesis, demonstrating the efficacy of centromere-mediated genetic engineering for inducing haploids in food crops with large genome sizes, as found in model plant *Arabidopsis*, and finally integrating DH technology into breeding pipelines beyond cereals and brassicas, such as bananas, cassava, and pulses.

1.8 Conclusions and Prospects

Haploids from their initial role in shaping the foundations of genetics to their current applications in diverse scientific disciplines have come a long way. Their simplicity and genetic tractability have made them powerful tools for understanding heredity, genetic disorders, and crop improvement. As technology advances the future of haploid research holds a great promise with potential breakthroughs in genetics, genomics, and agriculture. With an increasing number of successful applications, research on DHs has accelerated in the recent few decades (Forster et al. 2007). The number of newly discovered plant species reporting doubled haploidy has been steadily increasing. There were very few reports on the manufacture of DHs prior to the 1970s; nevertheless, this increased to approximately 50 publications in the 1970s, 185 in the 1980s, and approximately 200 in the 1990s (information from Maluszynski et al. 2003c). Applying DH technology to neglected crops is becoming quite popular, especially for high-value crops like pharmaceutical and aromatic species, where significant benefits can be made quickly. The adoption and adaption of procedures from model species, namely, tobacco, rapeseed, and barley, have played a significant role in the development of DH technologies for new species. Compared to the original procedures, published approaches currently include approximately 200 species and show a wide diversity of alterations. A technique that is optimized for one genotype is rarely also optimal for another. When adapting procedures from one species to another, significant adjustments are frequently needed, which are typically explained by empirical comparisons of protocols. Consequently, the number of laboratories working on creating new techniques is correlated with success. The multitude of variables that govern and impact androgenesis presents the biggest challenge to enhancing it. Even if it is successful, this will aid in comprehending only the initial stages of this intricate procedure. The same issues surround gynogenic techniques. Although there is usually no need for adjustments, the wide-crossing techniques are only applicable to a restricted set of species.

DH systems, while not flawless, are widely used in genetic mapping (Forster and Thomas 2005a, b), breeding (Thomas et al. 2003), and mutation studies (Szarejko and Forster 2007, Gajecka and Szarejko 2023). Additionally, they offer helpful platforms for basic research on plant embryogenesis and cell differentiation, as well as for genetic transformation. Further advancements in approaches will emanate as our understanding of the underlying mechanisms deepens. Collaborative efforts between breeders and scientists from other fields are required to determine the most effective methods for implementing DH systems in real-world applications. When combined with marker-assisted selection, DH technology is an effective tool for accelerating breeding and assisting in meeting concerns of climate change, food quality and higher crop productivity.

References

Abraha E, Klíma M, Vyvadilová M, Bechyne M (2008) Effectiveness of selected antimitotic agents in microspore culture for *Brassica carinata* doubled haploid production. Agric Trop Subtrop 41:137–149

Adams KL, Wendel JF (2005) Polyploidy and genome evolution in plants. Curr Opin Plant Biol 8:135–141

Allshire RC, Karpen GH (2008) Epigenetic regulation of centromeric chromatin: old dogs, new tricks? Nat Rev Genet 9:923–937

Almouslem AB, Jahuar PP, Peterson TS, Bomminieni VR, Rao MB (1998) Haploid durum wheat production via hybridization with maize. Crop Sci 38:1080–1087

Anderson SN, Johnson CS, Chesnut J, Jones DS, Khanday I, Wood-house M, Li C, Conrad LJ, Russell SD, Sundaresan V (2017) The zygotic transition is initiated in unicellular plant zygotes with asymmetric activation of parental genomes. Dev Cell 43:349–358

Aziz AN, Seabrook JEA, Tai GCC, De Jong H (1999) Screening diploid Solanum genotypes responsive to different anther culture conditions and ploidy assessment of anther derived roots and plantlets. Am J Patato Res 76:9–16

Bajaj YPS (1990) *In vitro* production of haploids and their use in cell genetics and plant breeding. Haploids in crop improvement 1:3–44. Springer, Berlin, Heidelberg. https://doi.org/10.1007/978-3-642-61499-6_1

Barclay IR (1975) High frequencies of haploid production in wheat (*Triticum aestivum*) by chromosome elimination. Nature 256:410–411

Barret P et al (2008) QTL control of androgenesis in maize anther culture. Plant Cell Rep 27:1147–1156

Batzenschlager M, Lermontova I, Schubert V, Fuchs J, Berr A, Koini MA, Houlne G, Herzog E, Rutten T, Alioua A, Fransz P, Schmit AC, Chaboute ME (2015) Arabidopsis MZT1 homologs GIP1 and GIP2 are essential for centromere architecture. Proc Natl Acad Sci USA 112:8656–8660

Begheyn RF, Vangsgaard K, Roulund N, Studer B (2016) In: Roldán-Ruiz I et al (eds) Breeding in a world of scarcity. Springer, pp 151–155

Bhojwani SS, Dantu PK (2010) Haploid plants. In: Davey MR, Anthony P (eds). Plant cell culture, pp 61–78

Bhowmik P, Bilichak A (2021) Advances in gene editing of haploid tissues in crops. Genes 12:1410

Blakeslee AF, Belling J, Farhnam ME, Bergner AD (1922) A haploid mutant in the Jimson weed, *Datura stramonium*. Science 55:646–647

Blasco M (2016) María Luisa Badenes and María del Mar Naval. Breed Sci 66:606–612

Bohanec B (2003) Ploidy determination using flow cytometry. In: Maluszynsky M, Kasha KJ, Forster BP, Szaejko I (eds) Doubled haploid production in crop plants: a manual. Kluwer Academic Publishers, Dordrecht, pp 397–403

Bordes J, Dumas de Vaulx R, Lapierre A, Pollacsek M (1997) Haplodiploidization of maize (*Zea mays* L.) through induced gynogenesis assisted by glossy markers and its use in breeding. Agronomie 17:291–297

Boutilier K, Offringa R, Sharma VK, Kieft H, Ouellet T, Zhang L, Hattori J, Liu CM, van Lammeren AA, Miki BL, Custers JB, van Lookeren Campagne MM (2002) Ectopic expression of BABY BOOM triggers a conversion from vegetative to embryonic growth. Plant Cell 14:1737–1749

Brazauskas G, Pasakinskiene I, Jahoor A (2004) AFLP analysis indicates no introgression of maize DNA in wheat maize crosses. Plant Breed 123:117–121

Brew-Appiah RAT, Ankrah N, Liu W, von Wettstein D, Rustgi S (2013) Generation of doubled haploid transgenic wheat lines by microspore transformation. PLoS One 11:e80155

Britt AB, Kuppu S (2016) Cenh3: an emerging player in haploid induction technology. Front Plant Sci 7:357. https://doi.org/10.3389/fpls.2016.00357

Burk LG, Gerstel DU, Wernsman EA (1979) Maternal haploids of Nicotiana tabacum L. from seed. Science 206:585

Burrack LS, Berman J (2012) Flexibility of centromere and kinetochore structures. Trends Genet 28:204–212

Caligari PDS, Powell W, Liddell K, de Maine MJ, Swan GEL (1988) Methods and strategies for detecting Solanum tuberosum dihaploids in interspecific crosses with S. phureja. Ann Appl Biol 112:323–328

Caperta AD, Delgado M, Ressurreição F, Meister A, Jones RN, Viegas W, Houben A (2006) Colchicine-induced polyploidization depends on tubulin polymerization in c-metaphase cells. Protoplasma 227:147. https://doi.org/10.1007/s00709-005-0137-z

Chalyk ST (1999) Use of maternal haploids for improving maize inbred lines. MNL 73:54–56

Chase SS (1949) Monoploid frequencies in a commercial double cross hybrid maize, and in its component single cross hybrids and inbred lines. Genetics 34:328–332

Chase SS (1969) Monoploids and monoploid-derivatives of maize (Zea mays L.). Bot Rev 35:117–167

Chaudhary HK (2007) Dynamics of doubled haploidy breeding and molecular cytogenetic approaches vis-à-vis genetic upgradation of bread wheat for organic and low input farming systems in north-western Himalayas. In: Proceedings of the EUCARPIA symposium on organic and sustainable, low-input agriculture with genotype x environment interactions, Wageningen, p 54

Chaudhary HK (2008a) Dynamics of wheat x *Imperata cylindrica*- a new chromosome elimination mediated system for efficient haploid induction in wheat. In: Appels R et al (eds) Proceedings of the 11th international wheat genetics symposium. University of Sydney Press, Sydney, pp 647–650

Chaudhary HK (2008b) Dynamics of doubled haploidy breeding and molecular cytogenetic approaches in bread wheat. In: Taniguchi K, Zhang X (eds) Focus on north-west Himalayan regions. The Society of Chromosome Research. Adv Chromosome Sci 3(2):67–69

Chaudhary HK (2010) Chromosome elimination process- a boon or bane for alien introgression in wheat. In: Zhang X et al (eds) Proc of the 4th Asian chromosome colloquium, Beijing, p 21

Chaudhary HK, Sethi GS, Singh S, Pratap A, Sharma S (2005) Efficient haploid induction in wheat by using pollen of *Imperata cylindrica*. Plant Breed 124(1):96–98

Chauhan H, Khurana P (2011) Use of doubled haploid technology for the development of stable drought tolerant bread wheat (Triticum aestivum L.) transgenics. Plant Biotechnol J 9:408–417

Chen K, Gao C (2013) TALENs: customizable molecular DNA scissors for genome engineering of plants. J Genet Genomics 40:271–279

Chen SJ, Song TM (2003) Identification haploid with high oil Xenia effect in maize. Acta Agronomica Sinica 29:587–590

Chen FQ, Prehn D, Hayes PM, Mulrooney D, Corey A, Vivar H (1994) Mapping genes for resistance to barley stripe rust (*Puccinia striiformis* f. sp. *hordei*). Theor Appl Genet 88:215–219

Chen A et al (2013) Advanced backcross QTL analysis of fiber strength and fineness in upland cotton. Mol Breed 32:409–418

Chen J, Strieder N, Krohn NG, Cyprys P, Sprunck S, Engelmann JC, Dresselhaus T (2017) Zygotic genome activation occurs shortly after fertilization in maize. Plant Cell 29:2106–2125

Cherkaoui S, Lamsaouri O, Chlyah A, Chlyah H (2000) Durum wheat x maize crosses for haploid wheat production: influence of parental genotypes and various experimental factors. Plant Breed 119:31–36

Choo TM (1981a) Doubled haploids for estimation of mean and variance of recombinants. Values Genet 97:165–172

Choo TM (1981b) Doubled haploids for studying the inheritance of quantitative characters. Genetics 99:525–540

Chugh A, Amundsen E, Eudes F (2009) Translocation of cell-penetrating peptides and delivery of their cargoes in triticale microspores. Plant Cell Rep 28:801–810

Clausen RE, Mann MC (1924) Inheritance in Nicotiana tabacum. V. The occurrence of haploid plants in interspecific progenies. Proc Nat Acad Sci 10:121–124

Coe EH (1959) A line of maize with high haploid frequency. Am Nat 93:381–382

Coe EH (1962) Spontaneous mutation of the aleurone color inhibitor in maize. Genetics 47:779–783

Colnago LA, Englesberg M, Souza AA, Barbosa LL (2007) High-throughput, non-destructive determination of oil content in intact seeds by continuous wave-free precession NMR. Anal Chem 79:1271–1274. https://doi.org/10.1021/ac062091. PMID: 17263365

Comai L (2005) The advantages and disadvantages of being polyploid. Nat Rev Genet 6:836–846

Comai L (2014) Genome elimination: translating basic research into a future tool for plant breeding. PLoS Biol 12:e1001876

Comai L, Tan EH (2019) Haploid induction and genome instability. Trends Genet 35:791–803

Conner JA, Mookkan M, Huo H, Chae K, Ozias-Akins P (2015) A parthenogenesis gene of apomict origin elicits embryo formation from unfertilized eggs in a sexual plant. Proc Natl Acad Sci USA 112:11205–11210

Conner J, Ozias-Akins P, Schmidt A (2017) Apomixis: engineering the ability to harness hybrid vigor in crop plants. Methods Mol Biol 1669:17–34

Cooper JL, Henikoff S (2004) Adaptive evolution of the histone fold domain in centromeric histones. Mol Biol Evol 21:1712–1718

Copenhaver GP, Preuss D (2010) Haploidy with histones. Nat Biotechnol 28:423–424

Cordts S, Bantin J, Wittich PE, Kranz E, Lorz H, Dresselhaus T (2001) ZmES genes encode peptides with structural homology to defensins and are specifically expressed in the female gametophyte of maize. Plant J 25:103–114

David JL, Dusautoir JC, Raynauld C, Roumet P (1999) Heritable variation in the ability to produce haploid embryos via pollination with maize and embryo rescue in durum wheat. Genome 42:338–342

De Maine MJ (2003) Potato haploid technologies. In: Maluszynski M, Kasha KJ, Forster BP, Szarejko I (eds) Doubled haploid production in crop plants – a manual. Kluwer Academic Publishers, Dordrecht/Boston/London, pp 241–247

Deimling S, Fleihinghaus-Roux T (1996) Haploids in rye. In: Mohan-Jain S, Spory SK, Veilleux RE (eds) In vitro haploid production in higher plants. Kluwer, Dordrecht, pp 181–204

Devaux P (2003) The Hordeum bulbosum (L.) method. In: Maluszynski M, Kasha KJ, Forster BP, Szarejko I (eds) Doubled haploid production in crop plants – a manual. Kluwer, Dordrecht/Boston/London, pp 15–20

Dong X, Xu X, Miao J, Li L, Zhang D, Mi X, Liu C, Tian X, Melchinger AE, Chen S (2013) Fine mapping of qhir1 influencing in vivo haploid induction in maize. Theor Appl Genet 126:1713–1720

Dong X, Xu XW, Li L, Liu CX, Tian XL, Chen SJ (2014) Marker-assisted selection and evaluation of high oil in vivo haploid inducers in maize. Mol Breed 34:1147–1158

Dong YQ, Zhao WX, Li XH, Liu XC, Gao NN, Huang JH, Wang WY, Xu XL, Tang ZH (2016) Androgenesis, gynogenesis, and parthenogenesis haploids in *Cucurbit* species. Plant Cell Rep 35:1991–2019. https://doi.org/10.1007/s00299-016-2018-7

Dunemann F, Schrader O, Budahn H, Houben A (2014) Characterization of centromeric histone H3 (CENH3) variants in cultivated and wild carrots (*Daucus* sp.). PLoS One 9:e98504

Dunwell JM (1985) Anther and ovary culture. In: Bright SWJ, Jones MGK (eds) Cereal tissue and cell culture. Nijhoff, Dordrecht, pp 1–44

Dunwell JM (2010) Haploids in flowering plants: origins and exploitation (review article). Plant Biotechnol J 8:377–424

Dwivedi SL, Britt AB, Tripathi L, Sharma S, Upadhyaya HD, Ortiz R (2015) Haploids: constraints and opportunities in plant breeding. Biotechnol Adv 33:812–829

Eder J, Chalyk S (2002) *In vivo* haploid induction in maize. Theor Appl Genet 104:703–708

El-Mounadi K, Morales-Floriano ML, Garcia-Ruiz H (2020) Principles, applications, and biosafety of plant genome editing using CRISPR-Cas9. Front Plant Sci 11:56. https://doi.org/10.3389/fpls.2020.00056

Eudes F, Shim Y-S, Jiang F (2014) Engineering the haploid genome of microspores. Biocatal Agric Biotechnol 3:20–23

Evans MM (2007) The indeterminate gametophyte1 gene of maize encodes a LOB domain protein required for embryo sac and leaf development. Plant Cell 19:46–62

Evtushenko EV, Elisafenko EA, Gatzkaya SS, Lipikhina YA, Houben A, Vershinin AV (2017) Conserved molecular structure of the centromeric histone CENH3 in *Secale* and its phylogenetic relationships. Sci Rep 7:17628

Ferrie AMR (2006) Doubled haploidy production in nutraceutical species-a review. Euphytica 158:347. https://doi.org/10.1007/s10681-006-9242-0

Ferrie AMR, Caswell KL (2011) Isolated microspore culture techniques and recent progress for haploid and doubled haploid plant production. Plant Cell Tissue Organ Cult 104:301–309. https://doi.org/10.1007/s11240-010-9800-y

Forster BP, Thomas WTB (2005a) Doubled haploids in genetics and plant breeding. Plant Breed Rev 25:57–88

Forster BP, Thomas WTB (2005b) In: Janick J (ed) Plant breeding reviews. John Wiley & Sons, pp 57–88

Forster BP, Heberle-Bors E, Kasha KJ, Touraev A (2007) The resurgence of haploids in higher plants. Trends Plant Sci 12(8):368–375. https://doi.org/10.1016/j.tplants.2007.06.007

Furusho M, Suenaga K, Nakajima K (1991) Production of haploid barley plants from barley x maize and barley x Italian ryegrass crosses. Jpn J Breed 41:175–179

Gaines EF, Aase HC (1926) A haploid wheat plant. Am J Bot 13:373–385

Gajecka M, Szarejko I (2023) Haploid mutagenesis: an old concept and new achievements. In: Penna S, Jain SM (eds) Mutation breeding for sustainable food production and climate resilience. Springer, Singapore. pp 129–150. https://doi.org/10.1007/978-981-16-9720-3_5

Gates RR (1929) A haploid Oenothera. Nature (London) 124:948

Geiger HH, Gordillo GA (2009) Doubled haploids in hybrid maize breeding. Maydica 54:485–499

Geiger HH, Braun MD, Gordillo GA, Koch S, Jesse J, Krützfeldt J (2006) Variation for female fertility among haploid maize lines. Maize Genet Coop News Lett 80:28–29

Germana MA (2003) Haploids and doubled haploids in citrus spp. In: Maluszynsky M, Kasha KJ, Forster BP, Szaejko I (eds) Doubled haploid production in crop plants: a manual. Kluwer Academic Publishers, Dordrecht, pp 303–308

Germana MA (2005) Protocol of somatic embryogenesis from Citrus spp. anther culture. In: Jain SM, Gupta PK (eds) Protocol of somatic embryogenesis-woody plants. Springer, Dordrecht, pp 191–207

Germana MA (2006) Doubled haploid production in fruit crops. Plant Cell Tissue Organ Cult 86:131–146

Germana MA (2009) Haploid and doubled haploids in fruit trees. In: Touraev A, Forster B, Jain M (eds) Advances in haploid production in higher plants. Springer, Heidelberg, pp 241–263

Germana MA (2011a) Anther culture for haploid and doubled haploid production. Plant Cell Tissue Organ Cult 104:283–300. https://doi.org/10.1007/s11240-010-9852-z

Germana MA (2011b) Gametic embryogenesis and haploid technology as valuable support to plant breeding. Plant Cell Rep 30:839–857

Gernand D, Rutten T, Varshney A, Rubtsova M, Prodanovic S, Bruss C, Kumlehn J, Matzk F, Houben A (2005) Uniparental chromosome elimination at mitosis and interphase in wheat and pearl millet crosses involvesmicronucleus formation, progressive heterochromatinization, and DNA fragmentation. Plant Cell 17:2431–2438

Gilles LM, Martinant JP, Rogowsky PM, Widiez T (2017) Haploid induction in plants. Curr Biol 27:R1095–R1097

Gohard FH, Zhiteneva AA, Earnshaw WC (2014) Centromeres. In: eLS. Wiley, Chichester. https://doi.org/10.1002/9780470015902.a0005785.pub2

Guha S, Maheshwari SC (1964) In vitro production of embryos from anther of *Datura*. Nature 204:497. https://doi.org/10.1038/204497a0

Guha S, Maheshwari SC (1966) Cell division and differentiation of embryos in the pollen grains of Datura *in vitro*. Nature 212(5057):27–28

Guo F, Huang BQ, Han Y, Zee SY (2004) Fertilization in maize indeterminate gametophyte1 mutant. Protoplasma 223:111–120

Gurushidze M, Hensel G, Heikel S, Schedel S, Valkov V, Kumlehn J (2014) True- breeding targeted gene knock-out in barley using designer TALE-nuclease in haploid cells. PLoS One 9:e92046

Hansen NJP, Andersen SB (1998) *In vitro* chromosome doubling with colchicine during microspore culture in wheat (*Triticum aestivum* L.). Euphytica 102:101–108

Häntzschel KR, Weber G (2010) Blockage of mitosis in maize root tips using colchicine- alternatives. Protoplasma 241:99–104

Hayes PM, Liu BH, Knapp SJ, Chen F, Jones B, Blake TK, Franckowiak JD, Rasmusson DC, Sorells M, Ullrich SE, Wesenberg D (1993) Quantitative trait locus effects and environmental interaction in a sample of North American barley germplasm. Theor Appl Genet 87:392–401

Henikoff S, Ahmad K, Malik HS (2001) The centromere paradox: stable inheritance with rapidly evolving DNA. Science 293:1098–1102

Henry Y (1998) Origin of microspore-derived dihaploid and polyhaploid *in vitro* plants. Plant Tissue Cult Biotechnol 4:127–135

Hermsen JGT, Ramanna MS (1981) Haploidy and plant breeding. Phil Trans R Soc Lond B 292:499–507

Hermsen JGT, Ramanna MS, Helsop-Harrison JGT, den Jijs APM (1981) Haploidy and plant breeding. Phil Trans R Soc Lond B 292:499–507

Heslop-Harrison JS, Schwarzacher T (2007) Domestication, genomics and the future for banana. Ann Bot 100:1073–1084

Heun M (1992) Mapping quantitative powdery mildew resistance of barley using a restriction fragment length polymorphism map. Genome 35:1019–1025

Hirsch CD, Wu YF, Yan HH, Jiang JM (2009) Lineage-specific adaptive evolution of the centromeric protein CENH3 in diploid and allotetraploid *Oryza* species. Mol Biol Evol 26:2877–2885

Hollingshead L (1928) A preliminary note on the occurrence of haploids in Crepls. Am Nat 62:282–284

Hooghvorst I, Nogués S (2020) Opportunities and challenges in doubled haploids and haploid inducer-mediated genome-editing systems in cucurbits. Agronomy 10(9):1441. https://doi.org/10.3390/agronomy10091441

Houben A, Sanei M, Pickering R (2011) Barley doubled-haploid production by uniparental chromosome elimination. Plant Cell Tissue Organ Cult 104:321–327

Hu G, Liang GH, Wassom CE (1991) Chemical induction of apomictic seed formation in maize. Euphytica 56:97–105

Huang BQ, Sheridan WF (1996) Embryo sac development in the maize indeterminate gametophyte1 mutant: abnormal nuclear behavior and defective microtubule organization. Plant Cell 8:1391–1407

Husbands A, Bell EM, Shuai B, Smith HM, Springer PS (2007) LATERAL ORGAN BOUNDARIES defines a newfamily of DNA-binding transcription factors and can interact with specific bHLH proteins. Nucleic Acids Res 35:6663–6671

Hussain T, Franks C (2019) In: Zhao Z-Y, Dahlberg J (eds) Sorghum, vol 1931. Springer, pp 49–59

Immonen S, Anttila H (1996) Success in rye anther culture. Votr Pflanzenzuchtg 35:237–244

Immonen S, Tenhola-Roininen T (2003) Protocol for rye anther culture. In: Maluszynsky M, Kasha KJ, Forster BP, Szaejko I (eds) Doubled haploid production in crop plants: a manual. Kluwer Academic Publishers, Dordrecht, pp 141–149

Inagaki M, Mujeeb-Kazi A (1995) Comparison of polyhaploid production frequencies in crosses of hexaploid wheat with maize, pearl millet and sorghum. Breed Sci 45:157–161

Inagaki MN, Pfeiffer WH, Mergoum M, Mujeeb-Kazi A (1998) Variation of the crossability of durum wheat with maize. Euphytica 104:17–23

Ingouff M, Rademacher S, Holec S, Soljic L, Xin N, Readshaw A, Foo SH, Lahouze B, Sprunck S, Berger F (2010) Zygotic resetting of the histone 3 variant repertoire participates in epigenetic reprogramming in *Arabidopsis*. Curr Biol 20:2137–2143

Ishii T, Ueda T, Tanaka H, Tsujimoto H (2010) Chromosome elimination by wide hybridization between Triticeae or oat plant and pearl millet: pearl millet chromosome dynamics in hybrid embryo cells. Chromosom Res 18:821–831

Ishii T, Karimi-Ashtiyani R, Houben A (2016) Haploidization via chromosome elimination: means and mechanisms. Annu Rev Plant Biol 67:421–438

Iwata A, Tek AL, Richard MM, Abernathy B, Fonseca A, Schmutz J, Chen NW, Thareau V, Magdelenat G, Li Y, Murata M, Pedrosa-Harand A, Geffroy V, Nagaki K, Jackson SA (2013) Identification and characterization of functional centromeres of the common bean. Plant J 76:47–60

Jacobsen E, Ramanna MS (1994) Production of monohaploids of Solanum tuberosum L. and their use in genetics, molecular biology and breeding. In: Bradshaw JE, Mackay GR (eds) Potato genetics. CAB International, Cambridge, pp 155–170

Jacquier NMA, Gilles LM, Pyott DE, Martinant JP, Rogowsky PM, Widiez T (2020) Puzzling out plant reproduction by haploid induction for innovations in plant breeding. Nat Plants 6:610–619

Jacquier NMA, Gilles LM, Pyott DE, Martinant JP, Rogowsky PM, Widiez T. (2020) Puzzling out plant reproduction by haploid induction for innovations in plant breeding. Nature Plants Perspective.

Jahne A, Becker D, Brettschneider R, Lorz H (1994) Regeneration of transgenic microspore derived fertile barley. Theor Appl Genet 89:523–533

Jain SM, Sopory SK, Veilleux RE (eds) (1996) Vitro haploid production in higher plants, vol 1, pp 11–33

Jauhar PP, Xu SS, Baenziger PS (2009) Haploidy in cultivated wheat's: induction and utility in basic and applied research. Crop Sci 49:737–755

Johannsen W (1903) Uber Erblichkeit in Populationen und in reinen Linien. Gustav Fischer, Jena

Johannsen W (1909) Elemente der exacten Erblichkeitslehre. Gustav Fischer, Jena

Johansen DA (1934) Haploids in hordeum vulgare. Proc Natl Acad Sci USA 20:98–100

Jorgensen GA (1928) The experimental formation of heteroploid plants in the genussolanum. J Genet 19:133–210. https://doi.org/10.1007/BF02984237

Jorgensen RB, von Bothmer R (1988) Haploids of Hordeum vulgare and H. marinum from crosses between the two species. Hereditas 108:207–212

Joubes J, Chevalier C (2000) Endoreduplication in higher plants. Plant Mol Biol 43:735–745

Jui PY, Choo TM, Ho KM, Konishi T, Martin RA (1997) Genetic analysis of a two-row × six-row cross of barley using doubled-haploid lines. Theor Appl Genet 94:549–556

Karimi-Ashtiyani R, Ishii T, Niessen M, Stein N, Heckmann S, Gurush-idze M, Banaei-Moghaddam AM, Fuchs J, Schubert V, Koch K, Weiss O, Demidov D, Schmidt K, Kumlehn J, Houben A (2015) Point mutation impairs centromeric CENH3 loading and induces haploid plants. Proc Natl Acad Sci USA 112:11211–11216

Kasha KJ (ed) (1974) Haploids in higher plants: advances and potential. Proceedings of the first international symposium, University of Guelph, Guelph, 10–14 June 1974

Kasha KJ (2005) Chromosome doubling and recovery of doubled haploid plants. In: Palmer CE, Keller WA, Kasha KJ (eds) Haploids in crop improvement II, vol 56. Springer, Berlin, pp 123–152

Kasha KJ, Kao KN (1970) High frequency haploid production in barley (Hordeum vulgare L.). Nature 225:874–876

Kasha KJ, Maluszynski M (2003) Production of doubled haploids in crop plants. An introduction. In: Maluszynski M, Kasha K, Forster BP, Szarejko I (eds) Doubled haploid production in crop plants. Springer, Berlin/Heidelberg, pp 1–4

Kasha KJ, Hu TC, Oro R, Simion E, Shim YS (2001) Nuclear fusion leads to chromosome doubling during mannitol pretreatment of barley (*Hordeum vulgare* L.) microspores. J Exp Bot 52:1227–1238
Kasha KJ, Shim YS, Simion E, Letarte J (2006) Haploid production and chromosome doubling. In: Fari MG, Holb I, Bisztray GD (eds) Acta horticulturae, vol 725. ISHS, Debrecen, pp 817–828
Katayama Y, Nei M (1964) Studies on the haploidy in higher plant. Rep Lab Plant Breed, Fac Agr, Miyazaki Univ, no 2: 1–78 (in Japanese)
Kato A (2002) Chromosome doubling of haploid maize seedlings using nitrous oxide gas at the flower primordial stage. Plant Breed 121:370–377
Kebede B (2020) Androgenesis in cereals: an overview. Ethiop J Sci Technol 13:35–46
Kebede AZ, Dhillon BS, Schipprack W, Araus JL, Bänziger M, Semagn K, Alvarado G, Melchinger AE (2011) Effect of source germplasm and season on the in vivo haploid induction rate in tropical maize. Euphytica 180:219–226
Kelliher T, Starr D, Wang W, McCuiston J, Zhong H, Nuccio ML, Martin B (2016) Maternal haploids are preferentially induced by CENH3-tailswap transgenic complementation in maize. Front Plant Sci 7:414
Kelliher T, Starr D, Richbourg L, Chintamanani S, Delzer B, Nuccio ML, Green J, Chen Z, McCuiston J, Wang W, Liebler T, Bullock P, Martin B (2017) MATRILINEAL, a sperm-specific phospholipase, triggers maize haploid induction. Nature 542:105–109
Kermicle JL (1969) Androgenesis conditioned by a mutation in maize. Science 166:1422–1424
Klíma M, Vyvadilová M, Kucera V (2008) Chromosome doubling effects of selected antimitotic agents in Brassica napus microspore culture. Czech J Genet Plant Breed 44:30–36
Kimber G, Riley R (1963) Haploid angiosperms. Bot Rev 29:480–531
Kloiber-Maitz M, Wieckhorst S, Bolduan C, Ouzunova M (2018) Haploidisierung in sorghum. European patent EP3366778A1
Knox RE, Clarke JM, De Pauw RM (2000) Dicamba and growth condition effects on doubled haploid production in durum wheat crossed with maize. Plant Breed 119:289–298
Kotyk JJ, Pagel MD, Deppermann KL, Colletti RF, Hoffman NG, Yannakakis EJ et al (2005) High- throughput determination of oil content in corn kernels using nuclear magnetic resonance imaging. J Am Oil Chem Soc 82:885–862. https://doi.org/10.1007/s11746-005-1155-5
Krohn NG, Lausser A, Juranic M, Dresselhaus T (2012) Egg cell signaling by the secreted peptide ZmEAL1 controls antipodal cell fate. Dev Cell 23:219–225
Kumlehn J (2016) Embryogenesis and plant regeneration from isolated wheat zygotes. In: Germana MA, Lambardi M (eds) In vitro embryogenesis in higher plants. Springer, New York, pp 503–514
Kumlehn J, Stein N (2014) Biotechnological approaches to barley improvement. Biotechnol Agric For 69:379–392. https://doi.org/10.1007/978-3-662-44406-1_20
Kumlehn J, Serazetdinova L, Hensel G, Becker D, Loerz H (2006) Genetic transformation of barley (Hordeum vulgare L.) via infection of androgenetic pollen cultures with Agrobacterium tumefaciens. Plant Biotechnol J 4:251–261
Kuppu S, Tan EH, Nguyen H, Rodgers A, Comai L, Chan SW, Britt AB (2015) Point mutations in centromeric histone induce post- zygotic incompatibility and uniparental inheritance. PLoS Genet 11:e1005494
Kynast RG, Davis DW, Phillips RL, Rines HW (2012) Gamete formation via meiotic nuclear restitution generates fertile amphiploid F1 (oat × maize) plants. Sex Plant Reprod 25:111–122
Lacadena L-R (1974) Spontaneous and induced parthenogenesis and androgenesis. In: Kasha J (ed) Khaploids in higher plants: advances and potential. Proceedings of the first international symposium, University of Guelph, Guelph, 10–14 June, pp 13–32
Lanaud C (2006a) Doubled haploids of cocoa (Theobroma cacao L.) 1. Observations of fertility. Plant Breed 99:187–195
Lanaud C (2006b) Doubled haploids of cocoa (Theobroma cacao L.) 2. Observations of monogenic and polygenic characters. Plant Breed 99:196–202
Lashermes P, Beckert M (1988) Genetic control of maternal haploidy in maize (*Zea mays* L.) and selection of haploid inducing lines. Theor Appl Genet 76:405–410

Laurie DA (1989) The frequency of fertilization in wheat × pearl millet crosses. Genome 32:1063–1067
Laurie DA, Bennett MD (1986) Wheat-maize hybridization. Can J Genet Cytol 28:313–316
Laurie DA, Bennett MD (1987) Wide crosses involving maize (*Zea mays*). Annual Report of the Plant Breeding Institute, 1986–87, pp 66
Laurie DA, Bennett MD (1989) The timing of chromosome elimination in hexaploid wheat-maize crosses. Genome 32:953–961
Laurie DA, Donoughe S, Bennett MD (1990) Wheat x maize and other wide sexual hybrids, their potential for genetic manipulation and crop improvement. In: Gustafson JP (ed) Genetic manipulation in plant development II. Plenum Press, New York, pp 95–106
Lermontova I, Kuhlmann M, Friedel S, Rutten T, Heckmann S, Sandmann M, Demidov D, Schubert V, Schubert I (2013) Arabidopsis KINETOCHORE NULL2 is an upstream component for centromeric histone H3 variant cenH3 deposition at centromeres. Plant Cell 25:3389–3404
Levan A (1945) A haploid sugar beet after colchicine treatment. Hereditas 31:399–410
Li L, Xu X, Jin W, Chen S (2009) Morphological and molecular evidences for DNA introgression in haploid induction via a high oil inducer CAUHOI in maize. Planta 230:367–376
Li X, Meng D, Chen S, Luo H, Zhang Q, Jin W, Yan J (2017a) Single nucleus sequencing reveals spermatid chromosome fragmentation as a possible cause of maize haploid induction. Nat Commun 8:991
Li X, Meng D, Chen S, Luo H, Zhang Q (2017b) Doubled haploid technology in maize breeding: mechanisms, applications, and perspectives. Mol Breed 37:135
Lichter R, Degroot E, Fiebig D, Schweiger R, Gland A (1988) Glucosinolates determined by HPLC in the seeds of microspore-derived homozygous lines of rapeseed (*Brassica napus* L.). Plant Breed 100:209–221
Lim W, Earle ED (2008) Effect of in vitro and in vivo colchicine treatments on pollen production and fruit set of melon plants obtained by pollination with irradiated pollen. Plant Cell Tissue Organ Cult 95:115–124
Lim W, Earle ED (2009) Enhanced recovery of doubled haploid lines from parthenogenetic plants of melon (Cucumis melo L.). Plant Cell Tissue Organ Cult 98:351–356
Lindstrom EW (1929) A haploid mutant in the tomato. J Hered 20:23–30
Lindstrom EW (1941) Genetic stability of haploid, diploid, and tetraploid genotypes in the tomato. Genetics 26:387–397
Lindstrom EW, Koos K (1931) Cyto-genetic investigations of a haploid tomato and its diploid and tetraploid progeny. Am J Bot 18:398–410
Liston A, Cronn R, Ashman TL (2014) Fragaria: a genus with deep historical roots and ripe for evolutionary and ecological insights. Am J Bot 101:1686–1699
Liu ST, Rattner JB, Jablonski SA, Yen TJ (2006) Mapping the assembly pathways that specify formation of the trilaminar kinetochore plates in human cells. J Cell Biol 175:41–53
Liu J, Guo TT, Wang HZ (2012) Development of automatic nuclear magnetic resonance screening system for haploid kernels in maize. Trans Chin Soc Agric Eng 28:233–236. https://doi.org/10.3969/j.issn.1002-6819.2012.z2.040
Liu D, Zhang H, Zhang L, Yuan Z, Hao M, Zheng Y (2014) Distant hybridization: a tool for interspecific manipulation of chromosomes. In: Pratap A, Kumar J (eds) Alien gene transfer in crop plants, vol 1 Innovations, methods and risk assessment. Springer, pp 25–42
Liu C, Li W, Zhong Y, Dong X, Hu H, Tian X, Wang L, Chen B, Chen C, Melchinger AE, Chen S (2015) Fine mapping of qhir8 affecting in vivo haploid induction in maize. Theor Appl Genet 128:2507–2515
Liu H et al (2017) High-efficiency CRISPR/Cas9 multiplex gene editing using the glycine tRNA-processing system-based strategy in maize. BMC Biotechnol 17:1–10
Liu C et al (2020a) Extension of the in vivo haploid induction system from diploid maize to hexaploid wheat. Plant Biotechnol J 18:316–318
Liu H et al (2020b) Efficient induction of haploid plants in wheat by editing of TaMTL using an optimized *Agrobacterium*-mediated CRISPR system. J Exp Bot 71:1337–1349

Lubberstedt T, Bohn M (2015) Competence center for doubled haploid research. XXIII Plant and, January 10–14. San Diago, CA, USA, p P0477
Lv H, Wang Q, Yang L, Fang Z, Liu Y, Zhuang M (2014) Breeding of cabbage (*Brassica oleracea* L. *var. capitata*) with fusarium wilt resistance based on microspore culture and marker-assisted selection. Euphytica 200:465–473
Magyar Z, De Veylder L, Atanassova A, Bako L, Inze D, Bogre L (2005) The role of the *Arabidopsis* E2FB transcription factor in regulating auxin dependent cell division. Plant Cell 17:2527–2541
Maheshwari S, Tan EH, West A, Franklin FC, Comai L, Chan SW (2015) Naturally occurring differences in CENH3 affect chromosome segregation in zygotic mitosis of hybrids. PLoS Genet 11:e1004970
Maluszynski M, Szarejko I, Sigurbjornsson B (1996) Haploidy and mutation techniques. In: Jain SM et al (eds) *In vitro* haploid production in higher plants, vol 1. Kluwer Academic Publishers, Dordrecht, pp 67–93
Maluszynski M et al (2003a) Published protocols for other crop plant species. In: Maluszynsky M, Kasha KJ, Forster BP, Szaejko I (eds) Doubled haploid production in crop plants: a manual. Kluwer Academic Publishers, Dordrecht, pp 309–336
Maluszynski M et al (2003b) Androgenesis and doubled haploid technology in crop plants. CAB Rev 8:1–29
Maluszynski M et al (2003c) Doubled haploid production in crop plants: a manual. Springer
Matzk F, Mahn A (1994) Improved techniques for haploid production in wheat using chromosome elimination. Plant Breed 113:125–129
Mayor PJ, Bernardo R (2009) Genomewide selection and marker-assisted recurrent selection in doubled haploid versus F2 populations. Crop Sci 49:1719–1725
Melchinger AE, Schipprack W, Würschum T, Chen SJ, Technow F (2013) Rapid and accurate identification of in vivo-induced haploid seeds based on oil content in maize. Sci Rep 3:2129. https://doi.org/10.1038/srep02129. PMID: 23820577
Melchinger AE, Schipprack W, Friedrich UH, Mirdita V (2014) In vivo haploid induction in maize: identification of haploid seeds by their oil content. Crop Sci 54:1497–1504. https://doi.org/10.2135/cropsci2013.12.0851
Mendiburu AO, Peloquin SJ, Mok DWS (1974) Potato breeding with haploids and 2n gametes. In: Kasha KJ (ed) Haploids in higher plants. University of Guelph, Guelph, pp 249–258
Michalik B, Adamus A, Nowak E (2000) Gynogenesis in polish onion cultivars. J Plant Physiol 156:211–216
Michelmore RW et al (1991) Identification of markers linked to disease resistance genes by bulked segregant analysis: a rapid method to detect markers in specific genomic regions by using segregating populations. Proc Natl Acad Sci USA 88:9828–9832
Mochida K, Tsujimoto H, Sasacuma T (2004) Confocal analysis of chromosome behavior in wheat × maize zygotes. Genome 47:224–228
Morinaga T, Fvkushima E (1931) Preliminary report on the haploid plant of rice, Oryza sativa L. Proc Imp Acad [Tokyo] 7:383–384
Mujeeb-Kazi A, Riera-Lizarazu O (1996) Polyhaploid production in the Triticeae by sexual hybridization. In: Jain SM, Sopory SK, Veilleux RE (eds) In vitro haploid production in higher plants, vol 1. Kluwer, Dordrecht, pp 275–296
Mukai Y, Okamoto G, Kiryu S, Takemoto S, Sharma SK, Yamamoto M (2015) The D-genome plays a critical role in the formation of haploid *Aegilops tauschii* through *Imperata cylindrica* mediated uniparental chromosome elimination. Nucleus 58:199–206
Muller HJ (1927) Artificial transmutation of the genes. Science 66:84–87
Murovec J, Bohanec B (2012) Haploids and doubled haploids in plant breeding. In: Abdurakhmonov I (ed) Plant breeding. InTech Europe, Croatia. ISBN: 978-953-307-932-5, pp 87–106
Nagaki K, Kashihara K, Murata M (2009) A centromeric DNA sequence colocalized with a centromere-specific histone H3 in tobacco. Chromosoma 118:249–257
Nagaki K, Terada K, Wakimoto M, Kashihara K, Murata M (2010) Centromere targeting of alien CENH3s in Arabidopsis and tobacco cells. Chromosom Res 18:203–211

Nagaki K, Yamamoto M, Yamaji N, Mukai Y, Murata M (2012) Chromosome dynamics visualized with an anti-centromeric histone h3 antibody in allium. PLoS One 7:e51315

Nair SK, Molenaar W, Melchinger AE, Boddupalli PM, Martinez L, Lopez LA, Chaikam V (2017) Dissection of a major QTL qhir1 conferring maternal haploid induction ability in maize. Theor Appl Genet 130:1113–1122

Nei M (1963) Efficiency of haploid method of plant breeding. Heredity 18:95–100

Neumann P, Navratilova A, Schroeder-Reiter E, Koblizkova A, Steinbauerova V, Chocholova E, Novak P, Wanner G, Macas J (2012) Stretching the rules: monocentric chromosomes with multiple centromere domains. PLoS Genet 8:e1002777

Newcomer H (1941) A colchicine-induced homozygous tomato obtained through doubling clonal haploids. Proc Am Soc Hortic Sci 38:610–612

Nordon A, Mcgill CA, Littlejohn CD (2001) Process NMR spectrometry. Analyst 126:260–272. https://doi.org/10.1039/B009293M. PMID: 11235115

Ohkawa Y, Suenaga K, Ogawa K (1992) Production of haploid wheat plants through pollination of sorghum pollen. Jpn J Breed 42:891–894

Op Den Camp RHM, Van Dijk PL, Gallard A (2017) Method for the production of haploid and subsequent doubled haploid plants. Patent WO 2017/200386 A1, 19 May 2017

Paz-Ares J, Ghosal D, Saedler H (1990) Molecular analysis of the C1-I allele from Zea mays: a dominant mutant of the regulatory C1 locus. EMBO J 9:315

Peloquin SJ, Gabert AC, Ortiz R (1996) Nature of "pollinator" effect in potato (Solanum tuberosum) haploid production. Ann Bot 77:539–542

Prasanna B, Chaikam V, Mahuku G (2012) Doubled haploid technology in maize breeding: theory and practice. CIMMYT

Prigge V, Xu X, Li L, Babu R, Chen S, Atlin GN, Melchinger AE (2012) New insights into the genetics of in vivo induction of maternal haploids, the backbone of doubled haploid technology in maize. Genetics 190:781–793. https://doi.org/10.1534/genetics.111.133066

Rao PS, Suprasanna P (1996) Methods to double haploid chromosome numbers. In: Jain SM, Sopory SK, Veilleux RE (eds) In vitro haploid production in higher plants, vol 1. Kluwer, Dordrecht, pp 317–339

Ravi M, Chan SWL (2010) Haploid plants produced by centromere-mediated genome elimination. Nature 464:615–618

Ren J, Wu P, Trampe B, Tian X, Lübberstedt T, Chen S (2017) Novel technologies in doubled haploid line development. Plant Biotechnol J 15:1361–1370

Riddle NC, Birchler JA (2003) Effects of reunited diverged regulatory hierarchies in allopolyploids and species hybrids. Trends Genet 19:597–600

Riera-Lizarazu O, Mujeeb-Kazi A, William MDHM (1992) Maize (Zea mays L.) mediated polyhaploid production in some Triticeae using a detached tiller method. J Genet Breed 46:335–346

Rietz S, Dermendjiev G, Oppermann E, Tafesse FG, Effendi Y, Holk A, Parker JE, Teige M, Scherer GF (2010) Roles of *Arabidopsis* patatin-related phospholipases a in root development are related to auxin responses and phosphate deficiency. Mol Plant 3:524–538

Rines HW (2003) Oat haploids from wide hybridization. In: Maluszynski M, Kasha KJ, Forster BP, Szarejko I (eds) Doubled haploid production in crop plants – a manual. Kluwer, Dordrecht/Boston/London, pp 155–160

Röber FK, Gordillo GA, Geiger HH (2005) In vivo haploid induction in maize – performance of new inducers and significance of doubled haploid line in hybrid breeding. Maydica 50:275–284

Roeber FK, Gordillo GA, Geiger HH (2005) In vivo haploid induction in maize. Performance of new inducers and significance of doubled haploid lines in hybrid breeding. Maydica 50(3):275–283

Rokka VM (2009) Potato haploids and breeding. In: Touraev A, Forster BP, Jain SM (eds) Advances in haploid production in higher plants. Springer Netherlands, Dordrecht, pp 199–208

Rokka VM, Valkonen JPT, Pehu E (1995) Production and characterization of haploids derived from somatic hybrids between Solanum brevidens and Solanum tuberosum through anther culture. Plant Sci 112:85–95

Ronceret A, Vielle-Calzada JP (2015) Meiosis, unreduced gametes, and parthenogenesis: implications for engineering clonal seed formation in crops. Plant Reprod 28:91–102
Saidi N, Chlyah O, Chlyach H (1998) Production of green durum wheat plants by pollination of wheat with maize. Can J Bot 76:652–656
Sanders ME, Franzke CJ (1962) Reduction of tetraploid sorghum to the diploid form after colchicine treatment. Nature 196:696–698
Sanei M, Pickering R, Kumke K, Nasuda S, Houben A (2011) Loss of centromeric histone H3 (CENH3) from centromeres precedes uniparental chromosome elimination in interspecific barley hybrids. Proc Natl Acad Sci USA 108:E498–E505
Sarkar KR, Pandey A, Gayen P, Mandan JK, Kumar R, Sachan JKS (1994) Stabilisation of high haploid inducer lines. MNL 68:64–65
Scherer GF, Ryu SB, Wang X, Matos AR, Heitz T (2010) Patatin-related phospholipase A: nomenclature, subfamilies and functions in plants. Trends Plant Sci 15:693–700
Seguí-Simarro JM, Nuez F (2007) Embryogenesis induction, callogenesis, and plant regeneration by in vitro culture of tomato isolated microspores and whole anthers. J Exp Bot 58:1119–1132
Segui-Simarro JM, Nuez F (2008) How microspores transform into haploid embryos: changes associated with embryogenesis induction and microspore-derived embryogenesis. Physiol Plant 134(1):1–12. https://doi.org/10.1111/j.1399-3054.2008.01113.x
Semiarti E, Ueno Y, Tsukaya H, Iwakawa H, Machida C, Machida Y (2001) The ASYMMETRIC LEAVES2 gene of *Arabidopsis thaliana* regulates formation of a symmetric lamina, establishment of venation and repression of meristem-related homeobox genes in leaves. Development 128:1771–1783
Shatskaya OA, Zabirova ER, Shcherbak VS, Chumak MV (1994) Mass induction of maternal haploids in corn. MNL 68:51
Shen Y, Pan G, Lubberstedt T (2015) Haploid strategies for functional validation of plant genes. Trends Biotechnol 33:611–620
Shewry PR (2009) Wheat. J Exp Bot 60:1537–1553
Simantel GM, Ross JG (1964) Colchicine-induced somatic chromosome reduction in sorghum. IV. An induced haploid mutant. J Hered 55:3–5
Singsit C (1991) Haploid induction in Mexican polyploid species and colchicine-doubled derivatives. Am J Potato Res 68:551–556
Snape JW (1976) A theoretical comparison of diploidized haploid and single seed descent populations. Heredity 36:275–277
Solmaz I, Sari N, Gursoy I, Kasapoglu S (2011) Comparison of *in vivo* and *in vitro* colchicines application for production of dihaploid 'Kirkagac' and 'Yuva Hasanbey' Melons. Afr J Biotechnol 10:15717–15724
Soltis PS, Marchant DB, Van de Peer Y, Soltis DE (2015) Polyploidy and genome evolution in plants. Curr Opin Genet Dev 35:119–125
Soriano M, Cistué L, Vallés MP, Castillo AM (2007) Effects of colchicine on anther and microspore culture of bread wheat. Plant Cell Tissue Organ Cult 91:225–234
Stoger E et al (1995) Plant transformation by particle bombardment of embryogenic pollen. Plant Cell Rep 14:273–278
Swanson WJ, Vacquier VD (2002) The rapid evolution of reproductive proteins. Nat Rev Genet 3:137–144
Szarejko I, Forster BP (2007) Doubled haploidy and induced mutation. Euphytica. https://doi.org/10.1007/s10681-006-9241-1
Tai GCC, Xiong XY (2003) Haploid production of potatoes by anther culture. In: Maluszynski M, Kasha KJ, Forster BP, Szarejko I (eds) Doubled haploid production in crop plants – a manual. Kluwer, Dordrecht/Boston/London, pp 229–234
Tek AL, Kashihara K, Murata M, Nagaki K (2010) Functional centromeres in soybean include two distinct tandem repeats and a retrotransposon. Chromosom Res 18:337–347
Thomas WTB, Forster BP, Gertsson B (2003) Doubled haploids in plant breeding. In: Maluszynsky M, Kasha KJ, Forster BP, Szaejko I (eds) Doubled haploid production in crop plants: a manual. Kluwer Academic Publishers, Dordrecht, pp 337–349

Thompson KF (1972) Oil-seed rape. In: Reports of the Plant Breeding Institute. Cambridge University Press, Cambridge, pp 94–96

Tian Y et al (2017) Precise base editing with a portable CRISPR/Cas12a guide RNA. Plant Biotechnol J 15:1915–1917

Tinker NA, Mather DE (1995) MQTL software for simplified composite interval mapping of QTL in multiple environments. J Agric Genom 1:1–5

Toojinda TE et al (1998) Introgression of quantitative trait loci (QTLs) determining stripe rust resistance in barley: an example of marker-assisted line development. Theor Appl Genet 96:123–131

Toole MG, Bamford R (1947) The formation of diploid plants from haploid peppers. J Hered 36:67–70

Touraev A et al (1997a) Plant male germ line transformation. Plant J 12:949–956

Touraev A, Vicente O, Heberle-Bors E (1997b) Initiation of microspore embryogenesis by stress. Trends Plant Sci 29:297–302

Touraev A, Pfosser M, Heberle-Bors E (2001) The microspore: a haploid multipurpose cell. Adv Bot Res 35:53–109. https://doi.org/10.1016/S0065-2296(01)35004-8

Touraev A, Forster BP, Jain SM (2009) Advances in haploid production in higher plants. Springer, Berlin, pp 1–208

Tuvesson S et al (2006) Molecular markers and doubled haploids in European plant breeding. Euphytica 158:305. https://doi.org/10.1007/s10681-006-9239-8

Tuvesson S, Dayteg C, Hagberg P, Manninen O, Tanhuanpaa P, Tenhola-Roininen T, Kiviharju E, Weyen J, Forster J, Schondelmaier J, Lafferty J, Marn M, Fleck A (2007) Molecular markers and doubled haploids in European plant breeding. Euphytica 158:305–312

Ushiyama T, Shimizu T, Kuwabara T (1991) High frequency of haploid production of wheat through intergeneric cross with teosinte. Jpn J Breed 41:353–357

Van Dan CMP, Lelivelt CLC, Movahedi S (2017) Non-transgenic haploid inducer lines in cucurbits. Patent WO 2017/081011 A1, 18 May 2017

Vijverberg K, Ozias-Akins P, Schranz ME (2019) Identifying and engineering genes for parthenogenesis in plants. Front Plant Sci 10:128

Wan Y, Petolino JF, Widholm JM (1989) Efficient production of doubled haploid plants through colchicines treatment of anther-derived maize callus. Theor Appl Genet 77:889–892

Wang M, Hoekstra S, Van Bergen S, Lamers GEM, Oppedijk BG, Van Der Heijden MW, De Priester W, Schilperoort RA (1999) Apoptosis in developing anthers and the role of ABA in this process during androgenesis in *Hordeum vulgare* L. Plant Mol Biol 39:489–501

Wang Z, Taramino G, Yang D, Liu G, Tingey SV, Miao GH, Wang LG (2001) Rice ESTs with disease- resistance gene or defense-response gene-like sequences mapped to regions containing major resistance genes or QTLs. Mol Gen Genomics 265:303–310

Wang G, He Q, Liu F, Cheng Z, Talbert PB, Jin W (2011) Characterization of CENH3 proteins and centromere-associated DNA sequences in diploid and allotetraploid *Brassica* species. Chromosoma 120:353–365

Wang H, Liu J, Xu X, Huang Q, Chen S, Yang P et al (2016) Fully-automated high-throughput NMR system for screening of haploid kernels of maize (corn) by measurement of oil content. PLoS One 11(7):e0159444. https://doi.org/10.1371/journal.pone.0159444

Watts A, Kumar V, Bhat SR (2016) Centromeric histone H3 protein: from basic study to plant breeding applications. J Plant Biochem Biotechnol 25:339–348

Wedzony M (2003) Protocol for doubled haploid production in hexaploid triticale (x Triticosecale Wittm.) by crosses with maize. In: Maluszynsky M, Kasha KJ, Forster BP, Szaejko I (eds) Doubled haploid production in crop plants: a manual. Kluwer Academic Publishers, Dordrecht, pp 135–140

Wędzony M, Marcińska I, Ponitka A, Ślusarkiewicz-Jarzina A, Woźna J (1998a) Production of doubled haploids in triticale (x Triticosecale Wittm.) by means of crosses with maize (Zea mays L.) using Picloram and Dicamba. Plant Breed 117:211–215

Wędzony M, Marciñska I, Ponitka A, Ślusarkiewicz-Jarzina A, Woźna J (1998b) Factors influencing triticale doubled haploid production by means of crosses with maize. In: Juskiw P (ed) Proceedings of the fourth international triticale symposium, vol 1, Lectures, Red Deer, 26–30 July 1998, pp 45–52
Wędzony M et al (2009) In: Touraev A et al (eds) Advances in haploid production in higher plants. Springer, pp 1–33
Weich EW, Levall MW (2003) Doubled haploid production of sugar beet (Beta vulgaris L.). In: Maluszynski M, Kasha K, Forster BP, Szarejko I (eds) Doubled haploid production in crop plants: a manual. Springer, Dordrecht, pp 255–263
Wendel JF, Richard CC (2003) Polyploidy and the evolutionary history of cotton. Adv Agron 78:139–186
Wijnker E, Vogelaar A, Dirks R, Van Dun K, De Snoo B, Van den Berg M, Lelivelt C, De Jong H, Chunting L (2007) Reverse breeding: reproduction of F1 hybrids by RNAi-induced asynaptic meiosis. Chromosom Res 15(2):87–88
Woo JW, Kim J, Kwon SI, Corvalan C, Cho SW, Kim H, Kim SG, Kim ST, Choe S, Kim JS (2015) DNA-free genome editing in plants with preassembled CRISPR-Cas9 ribonucleoproteins. Nat Biotechnol 33:1162–1164
Wu P, Li H, Ren J, Chen S (2014a) Mapping of maternal QTLs for in vivo haploid induction rate in maize (Zea mays L.). Euphytica 196:413–421
Wu P, Ren J, Li L, Chen S (2014b) Early spontaneous diploidization of maternal maize haploids generated by in vivo haploid induction. Euphytica 200:127–138
Wurschum T, Tucker MR, Reif JC, Maurer HP (2012) Improved efficiency of doubled haploid generation in hexaploid triticale by *in vitro* chromosome doubling. BMC Plant Biol 12:109
Yao L et al (2018) OsMATL mutation induces haploid seed formation in indica rice. Nat Plants 4:530–533
Yetisir H, Sari N (2003) A new method for haploid musk- melon (*Cucumis melo* L.) dihaploidization. Sci Hortic 98:277–283
Young ND (1994) Constructing a plant genetic linkage map with DNA markers. In: Ronald IKV, Phillips L (eds) DNA- based markers in plants. Kluwer, Dordrecht, pp 39–57
Yuan J, Guo X, Hu J, Lv Z, Han F (2015) Characterization of two CENH3 genes and their roles in wheat evolution. New Phytol 206:839–851
Zenkteler M, Nitzsche W (1984) Wide hybridization experiments in cereals. Theor Appl Genet 68:311–315
Zhang LQ, Yen Y, Zhang YL, Liu DC (2007) Meiotic restriction in emmer wheat is controlled by one or more nuclear genes that continue to function in derived lines. Sex Plant Reprod 20:159–166
Zhang J, Tang W, Huang Y, Niu X, Zhao Y, Han Y, Liu Y (2015) Down-regulation of a LBD-like gene, OsIG1, leads to occurrence of unusual double ovules and developmental abnormalities of various floral organs and megagametophyte in rice. J Exp Bot 66:99–112
Zhang W et al (2018) Efficient CRISPR/Cas9-mediated editing of trinucleotide repeat expansion in myotonic dystrophy patient-derived iPS and myogenic cells. Nucleic Acids Res 46:8275–8298
Zhang C et al (2020) Improving the efficiency of haploid induction in maize via machine learning. Gigascience 9:1–11
Zhao JP, Simmonds DH (1995) Application of trifluralin to embryogenic microspore cultures to generate doubled haploid plants in *Brassica napus*. Acta Physiol Plant 95:304–309
Zhao X, Xu X, Xie H, Chen S, Jin W (2013) Fertilization and uniparental chromosome elimination during crosses with maize haploid inducers. Plant Physiol 163:721–731
Zhao P, Begcy K, Dresselhaus T, Sun MX (2017) Does early embryogenesis in eudicots and monocots involve the same mechanism and molecular players? Plant Physiol 173:130–142
Zhong Y, Wang Y, Chen B, Liu J, Wang D, Li M, Qi X, Liu C, Boutilier K, Chen S (2022) Establishment of a dmp based maternal haploid induction system for polyploid *Brassica napus* and *Nicotiana tabacum*. J Integr Plant Biol 64(6):1281–1294. https://doi.org/10.1111/jipb.13244

Zhou WJ, Hagberg P, Tang GX (2002a) Increasing embryogenesis and doubling efficiency by immediate colchicine treatment of isolated microspores in spring *Brassica napus*. Euphytica 128:27–34

Zhou WJ, Tang GX, Hagberg P (2002b) Efficient production of doubled haploid plants by immediate colchicine treatment of isolated microspores in winter *Brassica napus*. Plant Growth Regul 37:185–192

In Vitro Culture of Haploids

2

Nirmala Bharti Patel and Zenu Jha

Abstract

Haploid cells have a gametophytic chromosomal number, and a doubled haploid results from chromosome doubling in haploid cells. Doubled haploid breeding can help advance traditional plant breeding programs and is useful in areas of genetics, genetic engineering, and plant breeding research. Though there are other ways to create doubled haploid, anther culture is the most popular and commonly used and is the successful method for the rapid recovery of pre-established breeding lines. Anther culture has been established in a wide range of plant species. Different factors such as plant growth conditions, physiology and age of donor plant, genotype, developmental stage of floral bud and anther, pre-treatment, culture medium composition, phytohormones, sucrose, and culture environment affect the anther culture efficiency and green plant regeneration. Genotype and pretreatment play a crucial role in the anther culture technology. This chapter examines the various haploid production methods, paying particular attention to in vitro anther culture methods of establishment and factors affecting anther culture response in plants.

Keywords

Haploids · In vitro culture · Anther culture · Pretreatment · Chromosome doubling · Flow cytometry

N. B. Patel
College of Agriculture & Research Station (CARS), Mahasamund, IGKV, Raipur, India

Z. Jha (✉)
Indira Gandhi Krishi Vishwavidyalaya, Raipur, India

Z. Jha et al. (eds.), *Doubled Haploids: Technological Advances and Role In Crop Improvement*, https://doi.org/10.1007/978-981-96-2339-6_2

2.1 Historical Account

Haploids can arise spontaneously without any known external influence, often appearing sporadically within a population or occasionally with a specific but low frequency. This frequency may be characteristic of the particular genetic line or species (Magoon and Khanna 1963). The initial observation of spontaneous haploid plant development was documented in *Datura stramonium* L. (Jimson weed) where two haploid individuals emerged among several plants displaying abnormal traits due to an attempt to induce chromosomal irregularities through cold exposure. Subsequently, the second confirmed haploid occurrence was noted in *Nicotiana tabacum*, followed by the first verified haploid instance in a cereal species, *Triticum compactum* var. *humboldtii*, a winter variety native to the Pacific Northwest region. In subsequent years, sporadic haploids have been documented in various species such as rice (Morinaga and Fukushima 1931; Ramiah et al. 1935), *Brassica* (Morinaga and Fukushima 1931; Fukushima and Iwasa 1964), and potato.

A methodology for rice androgenesis involving an indirect approach of utilizing callus induction and plant regeneration was detailed by Nishi and Mitsuoka (1969). An extensive investigation to determine the optimal developmental stage of microspores with the potential to transition into sporophytic forms was conducted by Chen (1977). It was previously documented that anthers from the Japonica type of rice alone demonstrated the ability to produce a satisfactory quantity of doubled haploids through anther culture, enabling predictable selection. In China, various rice varieties have been introduced using methods akin to those employed in wheat cultivar development, whereby anther culture post-hybridization is utilized to generate true breeding, self-pollinating cultivars. The case of "Hua Han Zhao" is particularly noteworthy as it involved two rounds of anther culture to amalgamate traits from multiple parental sources. Additionally, anther culture has been instrumental in producing inbred lines of rice that match or surpass the vigor and yield of their hybrid counterparts (Zhu and Pan 1990). Anther-derived rice cultivars are now extensively cultivated in China. Indica rice generally exhibits a low ability for anther culture (Dewi and Purwoko 2016), and this limitation is attributed to issues such as early anther necrosis, poor callusing capability, limited plant regeneration, and the frequent occurrence of albino plants. Research efforts to develop haploids of Indica rice through anther culture have also faced challenges, but there have been successful outcomes (Rout et al. 2016; Pusadkar and Jha 2018; Dash et al. 2022; Mayakaduwa and Silva 2023).

2.2 Techniques of Haploid Production

Long ago, the importance of haploids in the field of plant breeding and genetics was realized; however the limited frequency of their formation in nature has limited their practical application. The spontaneous natural production of haploids is caused by the process of apomixis or parthenogenesis. In vivo and in vitro approaches were used to develop haploids, but the success rate was much higher for in vitro approaches (Maluszynski et al. 2003). The in vivo techniques for haploid production include distant hybridization crosses followed by chromosome elimination and *bulbosum* technique (Germena 2010).

2.2.1 Distant Hybridization Method

The elimination of one of the parental genomes as a result of distant (interspecific or inter-generic) hybridization can result in the generation of hybrids. Some distant hybridization methods are used to produce haploids in wheat, including the Wheat × Maize Method (Maize Pollen Method), the Wheat × *Imperata cylindrica*, and the Wheat × Pearl Millet Method. The production of dihaploid plants in wheat has been found to be effective using these inter-generic crosses. Due to the preferential elimination of the pollen parent during embryo development, polyhaploid plants can be induced from such crosses, and haploid plants can be retrieved following embryo rescue (Kumar and Choudhary 2020).

2.2.2 *Bulbosum* Method of Haploid Production

In the early 1970s, interspecific hybridization of cultivated barley was reported. The gradual removal of *H. bulbosum* chromosomes from hybrid embryos between the two species results in the formation of haploid plantlets of *H. vulgare*. *H. bulbosum* is a perennial out-crossing species that is found in the Mediterranean region. The two cytotypes, diploid ($2n = 2x = 14$) and tetraploid ($2n = 4x = 28$), are used to generate haploid plants of barley. The haploid plant can further be converted into double haploid through treatment with colchicines (Kumar and Chaudhary 2020).

2.3 In Vitro Techniques for Haploid Production

In vitro induction of haploids is achieved either from male or female gametophytes. Additionally, it offers a different source for haploid generation in plants where male sterility predominates. However, the most significant limiting characteristics for rice and other cereal crops have been weak regeneration and low haploid efficiency. While haploid plants are naturally inclined to generate haploid offspring, the most effective and straightforward method to date for producing haploids or doubled haploids in a variety of species is in vitro anther culture or androgenesis.

2.3.1 Gynogenesis

The process of generating haploid embryos only from female gametophytes is known as gynogenesis. Following pollination, an egg cell may fertilize and develop into a hybrid embryo, in which case paternal chromosome elimination occurs during early embryogenesis or fertilization of the egg cell does not occur. Pollination of polar nuclei and the development of endosperm initiate the development of the haploid embryo. Ovarian or ovule cultures are used in gynogenesis, the in vitro process of inducing maternal haploidy. It is possible to trigger female gametophytes (megaspores) of angiosperms to develop into a sporophyte. The plants so produced are referred to as gynogenic haploids. The general procedure for gynogenic haploid production is briefly described. The flower buds are excised 24–48 hours prior to anthesis from un-pollinated ovaries. Following the excision of the calyx, corolla, and stamens, the ovaries undergo surface sterilization. The ovary is placed in the solid culture medium after having its end clipped at the distal portion of the pedicel. The ovaries are either placed on a filter paper or allowed to float above the media when using a liquid medium; the pedicel is inserted through the filter paper. The media that are most frequently employed are supplementary growth factors, MS, Whites, N6, and Nitsch. It is especially helpful to produce gynogenic haploids in plants with a male sterile genotype. This strategy outperforms another culturing method for such plant species.

Unfertilized ovules, ovaries, or flower buds can be cultured, depending on the species. Gynogenesis is the principal pathway leading to doubled haploid generation in certain members of the Cucurbitaceae, Lilliaceae, and Chenopodiaceae families. Gynogenetic haploids have been produced even in cases when anther or microspore culture is successful, such as in barley, maize, rice, and wheat. In many instances, embryogenic frequency is minimal; yet, in other instances, relatively high frequencies have been documented. The gynogenesis method's characteristics include the genetic stability of the doubled haploids and the lack of albinism (Touraev et al. 1996). Luthar and Bohanec (1999) and sugar beet (*Beta vulgaris*) have both successfully employed gynogenesis. Promising findings have been found about the frequency of haploid generation through gynogenesis in angiosperms, cucumbers, sweet potatoes, and some trees. Gynogenesis in sugar beet and onion is accomplished by culturing female gametophytes and then regenerating them. In cultured rice ovaries, apogamy of one of the synergids appeared to be the source of gynogenic embryos. In cultured ovaries, antipodals typically developed into multicellular glandular structures. The genotype of the donor cultivar determined the regeneration of gynogenic plants, with Japonica types being more responsive. It is advantageous to preserve the floral parts, and cultivating unhusked flowers that still had their pistil, stamens, and sterile glumes intact produced fruitful results. It has also helped increase sucrose from 1–2% to 3–6%. Ovarian culture produced 111 gynogenie rice plants, of which 77.5% were haploids and 22.5% diploids.

In actuality, it can be challenging to dissect unfertilized ovaries and ovules. Another drawback is that each flower only has one ovary. Anther/pollen cultures are employed more frequently in crop improvement programs to produce haploid plants than ovary/ovule cultures. On the other hand, one anther contains a significant number of microspores for culturing in vitro.

2.3.2 Androgenesis

Androgenic haploids are produced by anther or pollen culture. Pseudogamy and parthenogenesis are the processes used to create androgenic haploid embryos. The process of inducing and regenerating haploids and doubled haploids from male gametic cells is known as "in vitro androgenesis." Its great efficacy and broad applicability across many plant species make it highly promising for both commercial exploitation and plant breeding. More than 200 species from the Solanaceae, Cruciferae, and Gramineae families have been shown to regenerate from male gametes, although many woody and legume plants are fairly resistant (Wenzel and Foroughi-Wehr 1984; Germana 2010). Anther culture of Chinese rice varieties produced "Guan 18" and "Gan Xhao Xian No. 11," which mature in 105 days and have a high yield potential (5.6–6.0 t ha^{-1}) (Zhu and Pan 1990). Additionally, anther culture has produced a number of rice varieties in China (Tanfeng 1, Bua Yu 2, Hua Yu 1, Hua 03, Xin Xiu, Xhongua 8, Ta Be 78) and Argentina (Patei and Mocoi). The first DH lines released in India were "Satyakrishna" (CR Dhan 10) and "Phalguni" (CR Dhan 801) (CRRI Annual Report 2008, 2010).

Additionally, tobacco was the focus of groundbreaking research on the pretreatment effects of starvation, temperature shock, and/or other stress treatments (Touraev et al. 1996) as well as the genotypic control of androgenesis (Heberle-Bors 1984). In the field of fundamental biological research on cell division and the early phases of embryogenesis, the spring variety of rapeseed (*Brassica napus*) cv. "Topas" has emerged as a model genotype for dicotyledonous androgenesis (Binarova et al. 1993; Custers et al. 1994). Sood and Kumari (2019) used three different *Brassica campestris* varieties, HPBS-1, KBS-3, HPKM-04-1, HPBS-1 × HPKM-04-1, and KBS-3 × HPKM-04-1, as well as their cross combinations to determine the frequency of anther callus induction. The highest callusing in the androgenesis response was seen in the N6 medium containing HM2 (0.5 mg/l 2, 4-D + 1.0 mg/l NAA) and 3% sucrose. In general, the most promising genotype for inducing callus through anther culture was HPBS-1. In the future, haploids and double haploids can be successfully induced using this media and hormonal combination (Sood and Kumari 2019). In authors' laboratory, six rice varieties were studied by Chopkar et al. (2016): Indira Barani Dhan-1 (IBD-1), Karma Mahsuri, Swarna, MTU 1010, RF-75, and Samleshwari. The results showed that N6 media supplemented with 2,4-D hormone had a better response to callus induction, and MS media containing a combination of BAP, NAA, and Kinetin (3:1:1) favored higher plant regeneration, with more number of green plants.

2.3.3 Pollen (Microspore) Culture

Pollen mother cell goes through meiosis to create four haploid microspores (microsporogenesis) during normal pollen development. Microspores have a big vacuole and are uninucleated (have a single nucleus). Asymmetric microspore division, or microspore mitosis, produces two distinct cell types: a large vegetative cell and a small generative cell. The resultant bicellular product is known as pollen. The vegetative cell stops dividing in the G1 phase of the cell cycle, but the generative cell within the young pollen divides once more (second mitosis) to produce two sperm cells. At this point, starch granules and other storage products are accumulated by the pollen grain, causing it to completely enlarge. As a highly differentiated gametophytic cell, pollen at this stage is prepared to participate in the sexual reproduction process. Androgenic embryogenesis is the term for the process by which microspores produce asexual embryos, sometimes referred to as androgenic embryos. As a result, an entire reprogramming of the microspore's normal gametophytic development to sporophytic growth produces an androgenic embryo. Since the disruption at the first mitosis (microspore mitosis) is sufficient to push microspore toward the sporophytic pathway to produce embryos instead of pollen, several lines of experimental evidence suggest that the transition between microsporogenesis and microgametogenesis is a critical point to determine the pathway (Badiger 2017).

Microspores, or male gametophytic cells, or immature pollen can both result in haploid plants. Rather than going through a gametophytic pathway to differentiate into mature pollen grains or gametes, a male gametic cell, or microspore, produces embryos directly. Thus, the ability of a male gametic cell to entirely change from their gametophytic pathway of development to sporophytic growth and produce an embryo (direct embryogenesis) or through the formation of a callus (indirect embryogenesis) is known as androgenesis (Touraev et al. 1996). Before the microspores were eliminated, even in these early experiments, there was frequently a preculture phase for the buds or isolated anthers. The latter approach, which was first used on *Brassica napus*, has increased interest in the methodology for both fundamental and applied research (Heberle-Bors 1985; Huang 1992).

Using a glass rod to press and squeeze the anthers against a beaker's sides will extract the pollen. Debris from the anther tissue is filtered out of the pollen suspension. Large, viable pollen is concentrated through filtration, cleaned, and collected (smaller pollen does not regenerate). On either a solid or liquid medium, these pollens are cultivated. After the callus/embryo forms, it is transferred to an appropriate medium to eventually yield a haploid plant and, after that, a diploid plant (after treatment with colchicine). Improved tools and greater expertise are needed for isolated microspore culture. Important things to think about are the isolated microspore's viability and desiccation. The yields are not constant. Almost all of the microspores in the culture produced embryoids in certain instances, and none of them responded in other instances (Badiger 2017).

2.4 Anther Culture

One way to extract haploids from the pollen grains is through anther culture. Geneticists and plant breeders are particularly interested in them because they have a single set of chromosomes that can be used to create new crop cultivars. Anther cultures of several species in the families Brassicaceae, Poaceae, and Solanaceae have produced haploids. The process of producing androgenic haploids was hindered by Guha and Maheshwari's (1964) induction of haploid embryos from pollen

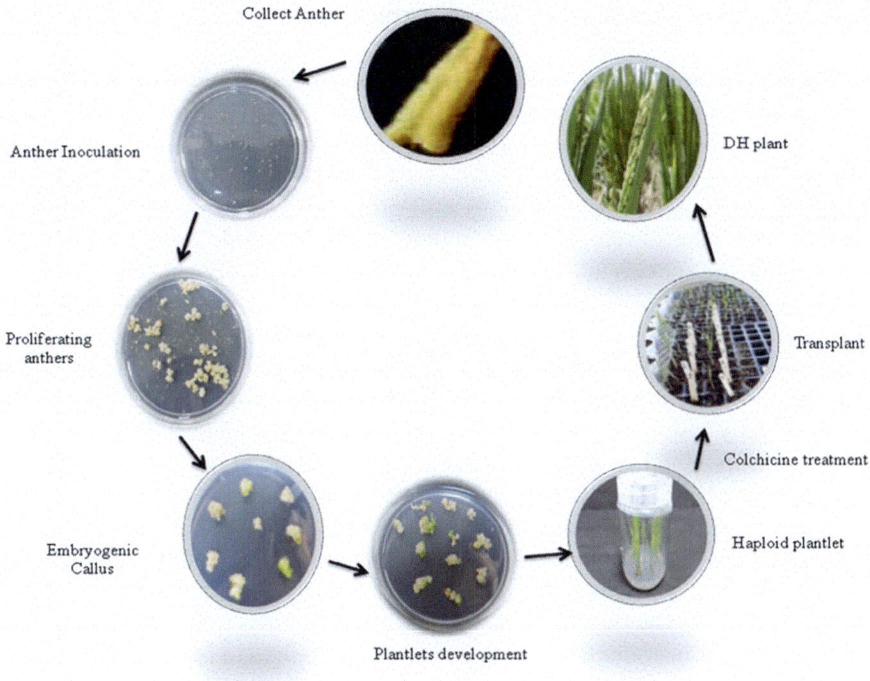

Fig. 2.1 Procedure of anther culture for the production of haploids and DHs of rice

grains. Bourgin and Nitsch (1967) were the first to obtain mature flowering plants of pollen origin from *N. tobacum* and *N. sylvestris*.

Anther culture is a very reproducible method for recovering homozygous plants with uncommonly desirable gene combinations and accelerating the breeding process in response to climate change. In the following sections, we have examined the different facets of anther culture in various crop species, with a focus on rice to improve genetics for agronomic double haploid breeding. Figure 2.1 outlines a step-by-step protocol for producing DH in rice.

There are several biological and culture environments' related factors which affect the anther culture response in plants. Several reports have presented screening of different factors and data on the optimized culture response. In the following sections, we describe different factors (see box) that influence anther culture response and efficiency of callus induction and plant regeneration.

- Plant growth conditions, physiology, and donor plant age.
- Genotype of the plant.
- Developmental stage of floral bud and anther.
- Surface sterilization.
- Pretreatment of cultures.
- Culture medium (composition, hormones, sucrose).
- Culture environment (light).

2.4.1 Stage of Microspore Development

A complex biological factor that has a significant impact on the success of anther culture is the pollen development stage. Generally speaking, the period of sensitivity to inductive treatments is around the first pollen mitosis or between the vacuolate microspore and early, mid-bicellular pollen. This is because the transcriptional status at that time is likely still proliferative and not fully differentiated. However, the developmental window of embryogenic competence varies depending on the species tested. Generally speaking, haploids can be produced by cultivating pollen at various stages of development (e.g., mature pollen, young-uninucleate, mid-uninucleate, late-uninucleate, binucleate, and pollen tetrad). Around the time of the first mitosis (uninucleate stage or early, mid-bicellular pollen grains), microspores typically show high androgenic competence. Pollen grains typically lose their ability to develop into embryos and follow the gametophytic developmental pathway once they start to build up storage reserves (Heberle-Bors 1985). Accurately determining the maturity stage of anther in Indica rice varieties (Silva and Ratnayake 2009), barley (Sriskandarajah et al. 2015), and chickpea buds can enhance anther culturing efficiency. Touraev et al. (1996) looked into whether the process of embryogenic induction in tobacco could be reversed. They came to the conclusion that stress treatments either starvation by itself or in conjunction with a heat shock applied to isolated microspores permanently prevented normal gametophytic development. When stressed tobacco microspores were transferred to a medium that contained more nucleosides, amino acids, and sucrose than the embryogenesis medium, they experienced a few cell divisions before degenerating instead of gametophytic pollen development continuing. As a result, the stress treatment reprogrammed the pollen to develop in a sporophytic manner rather than just stopping it. Pollen at the uninucleate stage produces mainly haploid plants, while pollen grains in some crops develop from late uninucleate to early bicellular stages (Germena 2010). When microspores are in the mid- to late uninucleate stage, which is typically reached when there are 3–6 cm separating the flag from the penultimate leaves, spikes are harvested. Various stresses applied in situ or in vitro have been shown to induce the switch of cultured microspores from a gametophyte to a sporophytic pathway.

The anthers with mid- to late uninucleate stages are identified by a cytological test of microspore using acetocarmine staining prior to culturing of anthers. Then, pollen-containing anthers from the mid-uninucleate developmental stage are cultured. Using the acetocarmine method, the proper stage for anther culture is determined by conducting tests on a few boots until the right appearance is achieved. The position of the anthers on spikelets has been used to subsequently determine the microspore of the anthers with mid- to late uninucleate stages (Naik et al. 2017). Rukmini et al. (2011) advise collecting panicles with mid- and late-stage microspores. Thus, it has been determined that the ideal stage for anther cultivation is in boots where the flag leaf and second leaf are 5 to 8 cm apart. In our study, the anthers from the panicles were separated, crushed in a glass slide, stained with

Fig. 2.2 Cytological microspore stages (Pusadkar and Jha 2018)

0.2 acetocarmine, and examined under a light field microspore in order to identify the stage of rice. The state of nudity is very obvious when it is less than 40×. The early microspore that emerges from the tetrad stage has an elliptical shape and is shrunken (Fig. 2.2). The germ pore and exile are visible, spherical, and swollen in the midmicrospore. Early bicellular pollen begins to exhibit generative and vegetative stages, late bicellular pollen begins to accumulate starch, and late microspores have an enlarged nucleus (Pusadkar and Jha 2018). Uninucleate microspores on the anthers produce haploids, whereas older and binucleate microspores produce diploids and triploids, respectively. Anther culture thus depends on harvesting the anthers at the appropriate time to obtain pollen with embryogenic competence. In other words, mature pollen grains are unsuitable for reliable sporophytic development because they cannot be effectively diverted from pollen tube formation, and only young microspores are productive for anther culture. Successful anther culture requires the selection of material at or close to the uninucleate stage of microspore development. Spikelet selection is based on anther location, size, and cytological observation. Although buds and spikes are chosen based on their size and morphological traits are useful as the first and preliminary indicators, the stage of microspore development needs to be verified under a microscope.

2.4.2 Pretreatment

Physical or chemical preculture treatments are applied to isolated flower buds, entire inflorescences, or anthers prior to culture to trigger the induction of sporophytic pathway, thereby preventing the development of fertile pollen (gametophytic pathway). According to Shariatpanahi et al. (2006), pretreatments like chilling, high temperature, high humidity, water stress, anaerobic treatment, centrifugation, sucrose and nitrogen starvation, ethanol, c-irradiation, microtubule disrupting agents, electrostimulation, high medium pH, and heavy metal treatment. Some of the methods which are popularly used in anther and in microspore culture are presented in Table 2.1. The most successful method for stimulating the development of pollen embryogenesis is the temperature shock. For instance, cold pretreatment (4 °C for 2–3 weeks) is frequently used in the anther culture of many crops, and it has genotype-dependent effects on three cultivars of spring rape (*Brassica napus* ssp. *oleifera*). For *Brassica* species, to effectively switch the developmental pathway, a brief high-temperature treatment (30–35 °C) is needed before further culture at 25 °C. Heat shock therapy could take the place of the starvation pretreatment, as demonstrated by Touraev et al. (1996). According to Vagera et al. (2004), the application of chemical and physical mutagens (e.g., gamma rays) to barley plants or to

Table 2.1 Protocols with special emphasis on pretreatment methods in anther culture in selected plant species

Plant	Stress	Protocol	Microspore stage	References
Wheat	Heat and starvation	33 °C and B medium/4 days/ anthers	Late UC premitotic	Touraev et al. (1996)
Brassica napus	Heat	32 °C/72 h/microspore suspension	Late UC	Custers et al. (1994)
Broccoli	Heat	32.5 °C/24 h/microspore suspension	Late UC-early BC	Davies and Morton (1998)
Barley	Cold	+4 °C/2–4 weeks/spikes	Early-mid-UC	Davies and Morton (1998)
Barley	Cold	+4 °C/3–4 days/anthers	UC	Jacquard et al. (2003)
Rice	Cold	8–10 °C/8 days/panicles	Mid-UC early BC	Zapata-Arias (1997)
Maize	Cold	+7 °C/1 week/tassels	Mid UC	Barnabas (1991)
Apple	Cold and starvation	4 °C/1–2 weeks/buds and then 4 °C and B med./2–3 days/ microspore suspension	Late UC	Hofer et al. (1999)
B. napus	Colchicine	Colchicine (25 mM)/42 h/25 °C/ microspore suspension	Late UC	Zhao et al. (1996)
Wheat	Colchicine and mild heat	Colchicine (0.04%)/29 °C/3 days/ anthers	Mid-late UC	Barnabas (1991)

UC unicellular, *BC* bicellular, *E* embryo, *R* regenerant, *C* callus, *IMC* isolated microspore culture, *AC* anther culture the seeds increased the androgenic response in recalcitrant rice varieties. Colchicine has been used to promote gametic embryogenesis and induce microspore division in a number of species, including rice, sugar beet, wheat, and chromosome doubling. Other pretreatments that are applied before anther culture include centrifugation, exposure to low air pressure, and water stress.

Buds that are to be pretreated are kept at a low temperature, sealed in a tiny plastic bag with a zipper, and wrapped in moist paper towel. It has been found that pretreating the anther after it has been removed from the plant and before it is incubated on culture medium can significantly boost anther culture response in a number of species. Members of the Solanaceae family provided the first evidence on this. Dunwell (2010) conducted an examination of the impact of a preculture period in a water-saturated atmosphere on the embryogenic potential of tobacco anthers. This method has been applied to other species, most notably cereals like barley, where treating the spikes for 28 days at 4 °C increases the yield of microspore calluses significantly. It is advised to treat potatoes (*Solanum tuberosum*) for 2 days at a temperature of 6.0 °C (Wenzel and Foroughi-Wehr 1984). In triticale, when it has been demonstrated that different preculture conditions may cause oxidative stress and alter cellular metabolic activity. These changes may then have an impact on cell viability and microspore embryogenesis efficiency. An effective pretreatment method is the application of mannitol, a sugar alcohol that is typically regarded as a non-metabolizable carbon source.

Studies by Chopkar et al. (2016) suggested that panicles can be wrapped in moist filter paper, placed in the dark for 7 to 10 days at 10 °C, and cleaned with distilled water and 70% ethanol. IBD-1 variety demonstrated the highest callus induction percentage (27.06%) and the lowest frequency were induced in rice variety Samleshwari (1.78%). Given that different genotypes react differently in vitro, the variable callus induction may be caused by genetic variations in the genotype composition. A preliminary investigation (Naik et al. 2017) was conducted to assess the callus induction potential over a period of 0–12 days at 4, 8, 10, and 12 °C. The results showed that, regardless of the combinations and concentrations of growth regulators in N6 basal media, the 10 °C temperature was the best. Anthers were incubated in the spikes for 7 days, and then they were exposed to 10 °C for 8 days, yielding 29.35% and 25.00% of the total. The rate of callus induction dramatically dropped to 7.17% with the extension of the incubation period from 8 to 10 days. Surprisingly, anthers incubated for 12 days had higher callusing rates (17.08%) than those incubated for 10 days (Naik et al. 2017).

From the discussion above, it is clear that that the selection of stress pretreatment will be based on factors such as overall viability, functionality, and practicality, as well as their effectiveness in the induction and frequency of embryogenic microspores. Furthermore, it is advisable to apply stresses as mildly as possible in order to reduce their adverse effects. In fact, stress is the primary factor that deviates microspores from their typical course of development. Microspores will always develop into pollen grains in the absence of stress. While some stresses appear to be useful in a limited number of species, others can be applied to a larger variety of species. For instance, heat and starvation, either separately or in combination, work well for *B. napus* (Brassicaceae), tobacco (Solanaceae), and wheat (Poaceae). Those stresses, however, are ineffective in tomatoes, which belong to the same family as tobacco. Furthermore, no embryos derived from *B. napus* microspores exist in *Arabidopsis*, a closely related species, although heat shock can easily induce embryogenesis in *B. napus* microspores. Cold treatment proved effective in attempts to produce doubled haploids in a crop species, as the others would need more handling and technical methods (Shariatpanahi et al. 2006). Cold pretreatment slows down the anther tissues' natural breakdown processes, shielding microspores from harmful substances released by the anthers as they decay. More embryogenic pollen grains are guaranteed to survive in the cold. Additionally, endo-reduplication occurs more frequently after cold pretreatment, increasing the number of spontaneously doubled haploid plants. Starvation results from the disconnection of microspores in cold-treated anthers from the tapetum (Sunderland 1971).

2.4.3 Surface Sterilization

Sterilization is required to eliminate surface contaminants (fungi and bacteria) prior to anther excision. To achieve anthers free of contaminants, numerous sterilization techniques have been employed. Following pretreatment, the flower buds are usually surface sterilized by immersing them for a short while in 70% (v/v) ethyl alcohol and then immersing them for another 10 to 15 min in a sodium hypochlorite solution (roughly 1.5% active chlorine in water) with a few drops of Tween 20. Finally, they are washed three times for 5 minutes each time using sterile distilled water. The final stage involves the aseptic removal of anthers from the filaments and their placement on the medium. Barley spikes are an exception, as they can only be sterilized by spraying 70% ethanol (Cistue et al. 2003) or by immersing in 70% ethanol for 5 min and then rinsing in sterile distilled water (Jacquard 2003). Under a stereoscopic microscope, anther extraction can be carried out in plants with tiny anther sizes, such as *Asparagus*, *Brassica*, *Trifolium*, and *Olea* (Bhojwani and Razdan 1983; Cistue et al. 2003). Usually, flower buds are thoroughly rinsed in sterile distilled water after being treated for 5 to 10 min with a 5% sodium or calcium hypochlorite solution to eradicate the infestation. Cut spikes with flag leaves from the inflorescence node of barley (Sriskandarajah et al. 2015) are stored in a beaker with a solution of 5.0 mg L^{-1} $CuSo_4.5H_2O$ and 62 gL^{-1} mannitol for 4 days at 4 °C in the dark. Following its removal from the spikes, the flag leaf is placed in a sterile Petri dish with a small amount (15 mm) of water added to maintain humidity levels.

In chickpea, buds of varying sizes are stored for 3 to 4 days in a plastic container and refrigerated at 4 °C (Grewal et al. 2009). After a 20-min soak in bleach, the anther was sterilized and cultured in a laminar air flow. The filament is carefully removed if the anther is still attached. If a solid medium is being used, the anthers are not being deeply embedded; rather, they are gently pressed onto the medium's surface. The anthers float on the surface when intact anthers are used in a liquid medium. The anthers from four to five buds (20–25 anthers) of a species with large anthers, like tobacco, can be cultured collectively on a single 100 × 15-mm-diameter Petri dish. Smaller Petri dishes or other containers might be more practical for species with smaller anthers or for specific experimental designs.

2.4.4 Media for Callus Initiation

A critical factor is the culture medium composition for its ability to induce microspore embryogenesis. Different crops exhibit differential requirements for basal media and other ingredients to promote the induction of callus and plant regeneration (Table 2.2). Compared to isolated microspores, the nutritional needs of the removed anthers are substantially less complex. Though there are many more, N6 medium (Chu 1978), modified MS medium (Murashige and Skoog 1962), Nitsch and Nitsch (1969) medium, and B5 medium (Gamborg et al. 1968) are the most often used basal media for anther culture. For Solanaceae, half-strength MS salt mixtures are generally advised, while N6 medium is recommended for cereals (Chu 1978).

Because of their osmotic and nutritional effects, the source of carbohydrates is necessary for the production of embryos in anther culture (Powell 1990). Since elevated concentrations of the carbon source appear to be harmful later in the culture period, the influence of carbohydrate concentration is likely related to regulation of osmotic pressure during the induction phase (Sunderland 1971). In plant

Table 2.2 Media used for anther callus induction in different crops

Crop	Media type	% callus	References
Niger	MS basal media supplemented with 2 mg/l 2,4-D and 0.3 mg/l KN	16% to 76%	Tesfaye et al. (2010)
Wheat	N6+ 2 mg/l 2,4-D+ 1 mg/l kinetin +90 g/l sucrose+ 7 g/l agar	0.67% to 18.00%	El-Hennawy et al. (2011)
Brassica campestris	N6+ 3% sucrose +0.5 mg/l 2,4-D+ 1.0 mg/l NAA+ 0.8 agar	78.02% to 59.69%	Sood and Kumari (2019)
Rice	N6+ 3 mg/l 2, 4-D+ 1 mg/l NAA+ 1 mg/l Kn.+ 0.8% w/v agar	0.02% to 5.68%	Gueye and Nadir (2010)
Rice	B5+ 1.0 mg 2,4-D/liter+ 0.5 mg IAA/liter+ 0.5 mg BAP/liter+ 0.8 agar	2.9% to 8.7%	Mohiuddin et al. (2014)
Rice	N6+ 5 mg/l 2, 4-D+ 1 mg/l NAA+ 0.5 mg/l kinetin+ $AgNO_3$+ 50 g/l maltose	0.75% to 5.15%	Bhattacharya et al. (2014)
Potato	MS+ 3.0 mg/l BAP+ 2.0 mg/l NAA	35.83% to 87.50%	Kumlay and Ercisli (2015)
Rice	N6+ 3% maltose+ 2.5 mg/l 2,4-D+ 0.8% agar	1.78% to 27.06%	Chopkar et al. (2016)
Rice	N6+ 3% maltose+ 2 mg/l 2,4-D+ 0.8% agar	1.17% to 3.53%	Minj et al. (2016)
Rice	N6+ 3% maltose+ 2 mg/l 2,4-D+ 0.8% agar	12.19% to 23.44%	Patel (2017)
Rice	N6+ 3% maltose+ 2 mg/l 2,4-D+ 0.8% agar	4.88% to 11.48%	Pusadkar and Jha (2018)

tissue, sucrose is the main translocated carbohydrate (Powell 1990). Lower levels, like 2–3%, are typically advantageous for species (like Solanaceae) where mature pollen is bicellular, but higher levels (6–17%) are necessary for those species (like Gramineae and Cruciferae) where mature pollen is shed in a tricellular condition. Due to its heat lability, sucrose, D-glucose, and D-fructose are combined in autoclaved media (Powell 1990). In barley anther culture, maltose has been effectively substituted for sucrose; typically, this is done at a concentration of 62 g/l in the induction medium and half this amount in the regeneration medium. Additionally, maltose has been added at concentrations ranging from 60 to 90 g/l to the anther culture medium of rice, rye, wheat, and triticale (Wedzony et al. 2009). In *Petunia* anther culture, it has been demonstrated that both glucose and fructose inhibit pollen embryogenesis (Raquin 1983). In clementine anther culture, lactose at 18 g/l and galactose at 9 g/l are frequently utilized (Germana 2010). In the anther culture of two clementine and two Mandarin cultivars, sucrose was found to be superior to glucose as a carbon source, whereas rye has shown glucose to be more stimulatory than sucrose (Wenzel and Foroughi-Wehr 1984). It was discovered that glycerol and sucrose together stimulated the development of calluses in clementines.

There have been reports of the addition of additional substances, including coconut water, glutamine, casein, proline, biotin, inositol, silver nitrate (ethylene antagonist), and polyvinylpyrrolidone (Powell 1990). Additionally, it has been discovered that adding exogenous aliphatic polyamines (PAs) to the culture medium increases the number of pollen-derived embryos in potato, cucumber, and clementine plants. Low-molecular-mass polycations called PAs—which include putrescine, cadaverine, spermidine, and spermine—are found in all living things and are categorized as growth regulators. They play a role in in vitro organogenesis and embryogenesis (Bagni and Tassoni 2001). Amino acids, which are associated with wall tissue deterioration, have been shown to increase induction rates in anther culture.

Usually, agar is added to anther culture media to solidify them. However, studies have shown that starch (potato, wheat, corn, or barley starch), gelrite, agarose, and ficoll can also be effective solidifying agents. Different responses have been observed in tests conducted on liquid, semisolid, and two-phase systems where anthers were floated on a liquid medium covering an agar-solidified medium (Dunwell 2010). A straightforward variation on anther culture is called "shed microspore" culture, in which anthers are encouraged to split open and release their microspores, typically into a highly osmolar liquid medium. It has been discovered that the technique works well with a number of species, including pepper and wheat.

The primary sources of variation during in vitro culture and regeneration are genotype and media composition. Bagheri and Jelodar (2008) found that there was a significant variation in callus induction based on genotype, media composition, and genotype × media interaction. El-Hennaway et al. (2011) assessed six genotypes and their F1 in wheat and found that all of them produced calli, with a range of 0.67% to 18.00% in the anthers that developed calli. The N6 induction medium of anther culture is supplemented with 90 g/L sucrose, 2 mg/L 2,4-D, 1 mg/L kinetin, and 7 g/L agar. In addition to providing nourishment, the culture medium controls the frequency of microspore callus induction and plays a crucial role in androgenesis. Since 1978, Heh-2, Heh-5, SK-3, SK-8, Szechuan medium, Chaleffs R-2, and Medium V have all been reported to induce N6-like properties. Because low inorganic ammonium nitrogen (NH4+) has been shown to be generally more beneficial for androgenesis in cereals, N6 medium with low NH4+ was developed and is most frequently used for rice anther culture (Chu et al. 1978). According to Silva and Ratnayake (2009), the induction of callus was almost three times greater on N6 medium compared to SK1 medium. Nevertheless, Indica strains demand even less NH4+. Raina and Zapata (1997) investigated the nitrogen needs of the Indica rice variety IR43 and created the MO19 medium, which is appropriate for Indica rice androgenesis. Indica rice varieties exhibit increased androgenic callus induction when N6 medium is supplemented with organic adjuvants such as coconut water, casein hydrolysate, and yeast extracts. The N6 medium (Chu 1978) and MS medium (Murashige and Skoog 1962) are solidifying agents used in callus induction experiments. They contain 3% maltose and 3% sucrose, respectively, and 2.5 mg/l 2,4-D. According to Dewi and Purwoko (2016), putrescine in the medium boosted the induction of callus, appropriately acted as a senescence inhibitor in the anther, preserved the viability of the anther, and extended the life of the anther cell wall to allow nutrients to reach the active microspore.

2.4.5 Culture Conditions

Although different culture conditions have been documented, anther cultures are typically incubated at 24–27°C and exposed to light at an intensity of roughly 2000 lux for 14 hours per 24 hour day. Vasil (1973), for instance, describes using alternating periods of darkness (12–6 h) at 22°C and light (12–18 h; 5000–10,000 lux/m^2). Nonetheless, each plant system must have its ideal conditions optimized (Bhojwani and Razdan 1983). It has occasionally been discovered that anthers must be continuously incubated in the dark. After the induction period, several species, including *Citrus clementina* Hort. ex Tan (Germana 2010), *Nicotiana tabacum* (Sunderland 1971), and *Hyoscyamus niger*, have also been demonstrated to benefit from an alternating light and dark period.

In anther culture, it has also been discovered that the density of the culture, that is, the quantity of anthers plated per volume of culture vessel or per unit volume of medium, and the technique of explant placement on the medium are crucial. It was suggested that when the cultures were exposed to dark at 23 ± 2 °C, callusing was observed in the Azucena rice variety (6.66%), but no callusing was found in the "Budda" variety. The rice varieties "Rajalaxmi and Ajay" were found to have the highest callus induction when anther culture was maintained at 25 ± 1 °C for 3–4 weeks (Mishra et al. 2011). Similarly, the highest regeneration frequency was obtained when an 18-hour photoperiod with a light intensity of 27 $\mu mol\ m^{-2}\ s^{-1}$ at 25 ± 1 °C was observed. The same conditions were also observed to play a role in a recalcitrant Indica variety ("RantaEmas," donor parent) and a Japonica variety ("Mankeumbyeo," recurrent parent). Minj et al. (2016) and Patel (2017) observed that culture in the dark favored anther culture response in two superior Indica rice hybrids: Safri-17 × PB 3 (Resistant to Blast) and Safri-17 × RYT3275.

2.4.6 Subculture

It is necessary to subculture on new media in order to preserve the cultures. At around 3 or 4 weeks after callus induction, the callus is subcultured by moving it or a part of it to a fresh medium. A large callus is divided into roughly equal-sized pieces and planted on various media in order to determine which media gave the best growth of the subcultured callus. In our haploid culture studies, anther callus derived from the cross between the wide-compatible Indica line T23 and the japonica line 668B was incubated for callus induction at 25 °C under dark conditions (Chopkar et al. 2016). Calli (2–3 mm) gradually appeared on the anther surface after about 25 days, at which point they were subcultured onto N6 medium (Chu 1978) supplemented with pH 5.8 sucrose (30 g/L), casein hydrolysate (0.3 g/L), NAA (1 mg/L), and 6-BA (3 mg/L). Subculture callus measuring 1–2 mm in diameter was placed in N6 regeneration media containing a solution of kietin (1, 1.5, 2 mg/l) and indol-6-acetic acid (0, 1, 2 mg/l). The callus was then incubated for 7 days in a sterile darkroom. Subsequently, it was exposed to light for 16 hours at 28 ± 2 °C and 1500–1600 lux until the plantlet regeneration was seen.

A study of Safri-17 × PB 3 (Resistant to Blast) and Safri-17 × RYT3275, two superior Indica rice hybrids, showed that cross Safri-17 × PB 3 generated callus, which was transferred to 16 different treatments (Minj et al. 2016) to determine the optimal media composition for converting callus for the regeneration of green plants (Fig. 2.3a). Hormonal combination of NAA/KIN/BAP with 0.1% colchicine in regeneration media was found to be the best. In our laboratory, we regularly use this optimized medium for green plant regeneration from anther cultures of Indica rice varieties (Fig. 2.3a, b).

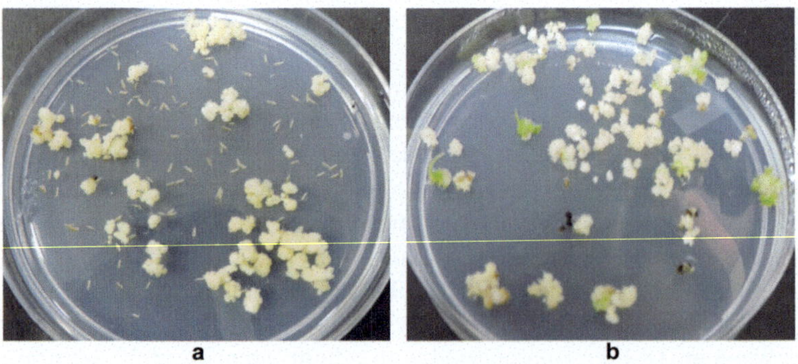

Fig. 2.3 (**a**) Plated anthers showing androgenic callus induction on N6 medium supplemented with maltose and NAA. (**b**) Greening and plantlets are developed on MS media containing growth hormones BAP, KIN, and NAA (Pusadkar and Jha 2018)

Fig. 2.4 Green plant regeneration on shoot induction medium (Pusadkar and Jha 2018)

2.4.7 Plant Regeneration

The main issue with rice culture is the production of albino plants, particularly in Indica rice varieties and hybrids (Raina and Zapata 1997) in which only 3% of Indica rice anther culture produced green plants (Zhang et al. 1989). According to Dewi and Purwoko (2016), *japonica* subspecies exhibits high anther culture ability and callus regeneration into green plants ranging from 20% to 60%. In our study (Chopkar et al. 2016), the highest percentage of green plant regeneration (0.66%) was found in varieties IBD-1 and Karma Mahsuri, while the remaining four varieties only displayed 0.33%. Less percentage of albino plants was found in the varieties IBD-1, MTU 1010, and Samleshwari (Chopkar et al. 2016). Cross Swarna sub 1 × IR 90019–17–159-5 and MTU1010 × Dagaddeshi had the highest percentages of green plants, at 21.21% and 3.15%, respectively (Patel 2017). Using Indica rice crosses, the percentage of green plants generated by anther culture were 43.75%, 12.5%, and MTU1010 × Dagaddeshi, variety RF-75, and Swarna sub 1 × IR 90019-17-159-5, respectively (Tekam 2017). Green plant regeneration involved the use of growth hormone composition such as NAA, KIN, and BAP (Fig. 2.4). The regeneration frequency was increased to 12.5% when IAA and kinetin were added to regeneration media. Consequently, in order to create plantlets, the embryogenic callus of a black rice cultivar was harvested and allowed to regenerate in N6 medium that was supplemented with varying amounts of IAA and kinetin (Pusadkar and Jha 2018).

2.4.8 Rhizogenesis

Regenerated plants are induced for rooting on nutrient media for the purpose of allowing good root growth and enabling survival in greenhouse conditions. In our laboratory, our studies (Minj et al. 2016, Pusadkar and Jha 2018) showed that for vigorous root development, green shoots measuring 2–3 cm in length can be moved

Fig. 2.5 Hardening of regenerated plants: (**a**) hydroponics, (**b**) plastic tray with soil, and (**c**) transferred in the field (Pusadkar and Jha 2018)

to MS medium supplemented with kinetin (0.25 mg) and NAA (1.00 mg) followed by another 4–6 weeks of culture on MS medium without hormones to promote root development (Fig. 2.5). The well-rooted regenerated plants are moved to a greenhouse where they are grown until they reached maturity.

2.4.9 Hardening/Acclimatization

Acclimatization ensures step-by-step transfer from the hydroponic setup to the net house and field conditions (Fig. 2.5a–c). Necessary culture management is implemented to ensure the plants' healthy growth and development. Plants are screened for variability in height, panicle size, and grain types. In our laboratory, after being transplanted to a plastic tray (30 × 45 cm), the DH plantlets are placed in 72 pots filled with clay soil and left to acclimate for 1 month under 50% shade (Minj et al. 2016). Mist spraying is used to supply water for irrigation. A field of 4 meters, with rows of 20 × 20 centimeters, is used for planting of 100 acclimated plants directly onto clay soil. Well-developed young rice plants with robust roots are moved to the greenhouse in pots filled with sterilized soil after 7 to 14 days. Plants that grow to a height of 8 to 10 cm are moved into bigger pots and subsequently to the field (Fig. 2.5c).

2.4.10 Diploidization Through Chromosome Doubling

Following anther culture, haploid plants can develop normally in vitro and can be acclimated to become healthy, mature plants. These plants frequently exhibit decreased vigor. Over several decades, numerous techniques have been used and are still being developed (Rao and Suprasanna 1992; Castillo et al. 2009). The most commonly used is in anti-microtubule agent, colchicine. The French chemists P.S. Pelletier and J. Caventou isolated it in 1820, and it is an alkaloid. Harvested from the seeds and corn of plants in the genus *Colchicum* (*Colchicum autumnale*, also called "meadow saffron"), colchicine is a toxic alkaloid and a secondary

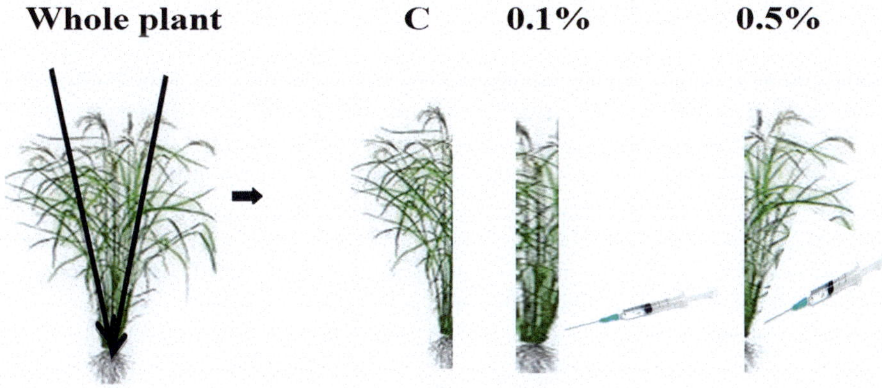

Fig. 2.6 Colchicine treatment of plants (Pusadkar and Jha 2018)

metabolite. Our observations suggest that plant sections treated with the suitable media showed proper chromosome replication; however, spindle formation was inhibited, and the cytoplasmic phase of cell division did not occur. All plants grown from anther cultures receive a colchicine treatment to double their chromosome count during the tillering stage. The plant was split into three sections and given treatment, 100 mg and 500 mg of colchicine, and a control portion that received no treatment (Fig. 2.6a–c). Plants receiving the colchicine dosage treatment in their tillers were able to double their chromosome count. Plants treated with colchicine are replanted in the field after a 10-minute break. Out of the 963 plants that were produced in the lab, 179 plants that are treated with different colchicine treatments yielded seeds when transferred to field conditions for secondary hardening. Each plant's seeds are collected in a separate envelope containing seeds from various plant treatments with different amounts of colchicine (0.1 mg/l, 0.5 mg/l, and control). Additionally, 69 plants out of 124 that received the same treatment in D × R crosses have produced seeds. Out of 42 plants in the cross S × P, 6 plants yielded seeds (Pusadkar and Jha 2018).

Other Chromosome Doubling Agents

Herbicides that are effective at micromolar concentrations include oryzalin, amiprophosmethyl (APM), trifluralin, pronamide, acenaphthene, chloramphenicol, nitrous oxide, parafluorophenylalanine, and hydroxyquinone (Rao and Suprasanna 1992; Castillo et al. 2009). Anti-microtubule medications can be added to microspore pretreatment media or used at different stages of androgenesis. Nitrogen oxide (N_2O) treatment is an exception that was created for maize seedlings. Plantlets at the six-leaf stage, during which time the plants develop flower primordia, are treated for 2 days at a high pressure of 600 kPa. The mechanism of action was suggested to be due to depolymerization of microtubules.

2.4.11 Detection of Haploids

In addition to producing only haploid plants, anther culture can also result in the plantlets of different ploidy (diploid, triploid, tetraploid, etc.) (D'Amato 1977; Dunwell 2010). In order to assess the regenerants and distinguish the desired non-haploid regenerants from haploid/DHs, it is crucial to ascertain the ploidy level. Breeders employ more indirect methods in the past to assess ploidy levels, such as comparing the morphology (height, width, and form of leaves, flowers, etc.) of donor and regenerated plants, as well as their vigor and fertility. These methods are simple but often unreliable, since environmental factors can affect the markers used to measure ploidy levels. Early on, isozyme analysis, progeny resting after self-pollination, and phenotypic markers were the main methods used to evaluate regenerated plants. Recent advances in biotechnology have made it possible to test homozygosity and determine the origin of plants more quickly and accurately (Badiger 2017). The haploids can be detected by the using of following techniques.

- *Stomata-Based Technique.*

We now know that in maize, haploids have smaller cells and less vegetative and floral parts than diploids. Out of 10,000 seedlings, Greenblatt and Bock (1967) chose 45 on the basis of size (e.g., thin coleoptiles, etc.), and five of them turned out to be monoploids, indicating some success in the haploid screening process. Additional factors to aid in preliminary screening include stomata size, pollen size, and pollen abortion scan; however, cytological confirmation is necessary in each instance. Stomata's primary job is to open and close in order to control the rates at which carbon dioxide is absorbed and water is lost. The diffusion of carbon dioxide and water vapor is impeded by stomata. The resistance to gas exchange is infinite when stomata are closed. Put differently, stomata function as a strong barrier to prevent carbon dioxide and water vapor from entering and leaving the leaf. Carbon dioxide and water vapor exchange occurs when stomata are open.

The observation that the number of chloroplasts in stomatal guard cell size increases in response to increased ploidy state forms the basis for determining ploidy using chloroplast characteristics. The plant ploidy level can occasionally be determined by measuring stomatal length (Thomas et al. 1975). In order to distinguish between haploid and diploid plants of *Actinidia deliciosa*, it was reported that stomatal length was a suitable criterion (Przywara et al. 1988). After determining that the mean stomatal length of haploid plants was 24 ± 1.7 μm and for diploid plants it was 33 ± 2.4 μm, the author concluded that measuring stomatal length is a quick way to determine the ploidy level of kiwifruit. However, no research has been done on sweet pepper. Using anther culture-derived plants from seven cultivars, Derby, Special, Bossanova, Boggie, and Minipaprika, Shrestha and Kang (2016) observed the suitability of the stomatal length and density as a criterion in the distinction between haploid ($2n$ = 12) and diploid ($2n$ = 24) plants of Sweet Pepper (*Capsicum annuum* L.). Plants were cultivated in screen houses, and measurements of a plant's length, width, and stomata impression were taken from at least ten fully grown leaves. Under a 1000× microscope, the length and number of stomata were noted. The findings indicated that the mean stomata length and number in haploid were 26.4 ± 2.4 μm and 7.4 ± 1.80 μm, respectively, while in diploid they were 35.2 ± 2.5 μm and 5.8 ± 0.8 μm or 33.3% and 30% higher in diploid than in haploid. Similarly, the average leaf length and width in haploid were 7.2 ± 1.2 cm and 3.98 ± 0.97 cm, respectively, whereas in diploid they were 9.66 ± 1.7 cm and 6.3 ± 1.0 cm, which indicated a 34.2% and 65.8% increase in diploid compared to haploid. Therefore, these parameters showed a significant difference between haploid and diploid plants, and measuring stomata length is a quick way to determine a plant's ploidy level in sweet peppers.

Additional information on the distribution of male gametophytic ploidy and the corresponding stability of the reproductive program can also be obtained from pollen size (Zlesak 2009). Pollen size is frequently regarded as an appealing parameter in plant research to track the mode of male reproduction and evaluate the fertility and/or ploidy of pollen at anthesis. Typically, image-based methods are used to measure pollen size in one or two dimensions, registering either the diameter or the transsectional particle area. This methodology presents three primary drawbacks. At first, the associated staining procedure and slide preparation protocol often change the morphometric features of the pollen grains and alter the size and morphology of the mature pollen, leading to aberrations in size determination. Second, one- and two-dimensional pollen size measurements typically provide an inaccurate and/or simplified representation of the actual pollen grain size because pollen from many plant species has a distinct, non-spherical body shape in addition to the presence of pores and exine ornaments. Third, image-based size approaches frequently exhibit errors in particle size estimation; primarily due to technical problems like out-of-focus or improper image analysis settings (e.g., debris inclusion).

Chromosome Counting

This is one of the accurate ways to assess ploidy levels. During mitotic or meiotic cell division, chromosome counting can be used to identify DH plants and haploid plants. Using root tips or other meristematic tissues, chromosomal counting during mitotic division is a simple and quick process (Maluszynska 2003). However, counting mitotic chromosomes to determine ploidy is a laborious and challenging process, particularly for crops like *Brassica napus* that have small chromosomes (Weber et al. 2004). Chromosome counting in the apical root meristems stained with DAPI (Axio Imager A2 microscope with fluorescence) was another method used to assess the ploidy level of plants. The number of chromosomes and the number of chloroplasts in the turnip leaf guard cells were found to be related, as demonstrated by chromosome counting in meristem root-tip cells stained with DAPI. Rice plants that were diploid or double haploid had 20 chromosomes, whereas plants that were haploid had 10 chromosomes.

Flow Cytometer-Based Estimation

A flow cytometer is used to examine moving particles in a suspension. These produce an epifluorescence when they are stimulated by a light source, such as a laser

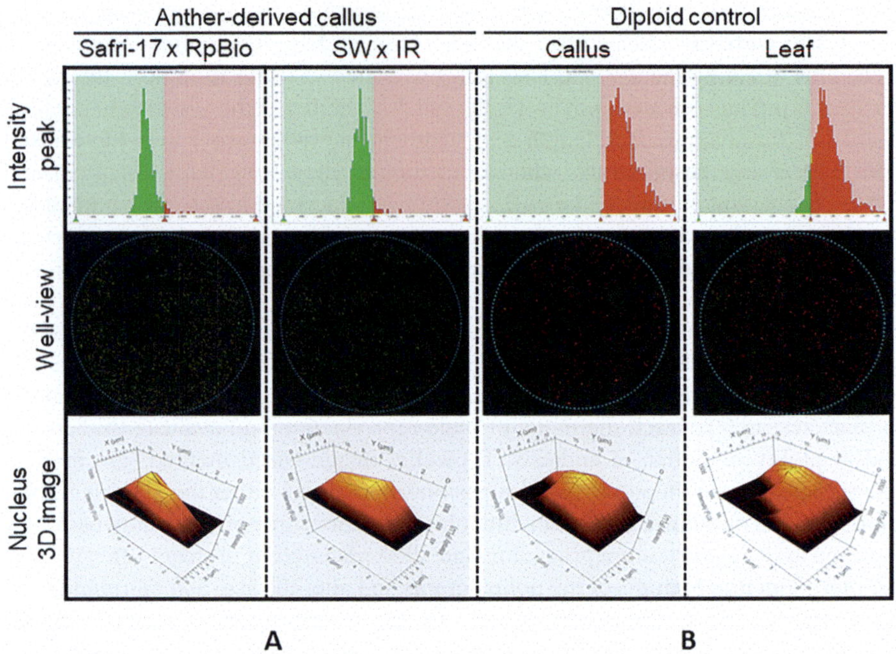

Fig. 2.7 FL-3 DNA intensity peak, the well view showing haploid cells in green and diploid cells in red, and the 3-D image of haploid and diploid nucleus in anther-derived haploid callus (**a**) and diploid control (**b**) (Pusadkar and Jha 2018)

or UV light (Ochatt 2008; Sahoo et al. 2019). The equipment's program translates these signals into a graph that shows the number of cells emitting epi-fluorescence at a particular time against the intensity of the fluorescence that is released. The staining solutions used to stain leaf or callus samples are propidium iodide and DAPI. Nuclear DNA content can be measured quickly and accurately with flow cytometry, which uses DNA selective fluorochromes. When plantlets emerge from microspore culture, the ploidy level can be measured using flow cytometry at an early developmental stage. Thus, it offers a quick and easy way to determine ploidy on a large scale in the early stages in Fig. 2.7.

2.5 Advantages of DHs

Using a variety of DH techniques in multiple crops, more than 280 varieties have been created; most of the protocols are known as anther culture. For instance, it's estimated that 50% of the barley (*Hordeum vulgare*) cultivars that are currently available in Europe were produced using a DH system, while in Canada, three of the five most popular cultivars across all grades of the Canada Western Red Spring (CWRS) wheat class in 2007 were DH cultivars (Dunwell 2010). Plant breeders can save a great deal of time by using a single round of recombination to produce homozygous lines. Additional benefits include the creation of numerous homozygous lines, effective genetic analysis, and the rapid development of useful trait markers. More specialized advantages include the potential for seed propagation as a substitute for vegetative multiplication in ornamentals, and doubled haploidy offers new options for traditional breeding methods in species like trees where long life cycles and inbreeding depression prohibit them (Forster and Thomas 2003).

Additionally, homozygous DHs open up new avenues for woody plant breeding and genetic research. It is impossible to create homozygous breeding lines for tree species using traditional techniques that require multiple generations of selfing. These species are typically distinguished by a protracted reproductive cycle, a high level of heterozygosity, size, and, occasionally, self-incompatibility (Germana 2010). Furthermore, haploid and homozygous plants shrinking in size relative to diploid and heterozygous plants may be useful in horticulture for dwarfing rootstocks for fruit crops or decorative plants, among other uses. The "gametoclonal variation," which refers to variations in morphological and biochemical traits as well as in chromosome number and structures that are observed among plants regenerated from cultured gametic cells, provides another chance to use haploids in crop improvement (Evans et al. 1984; Morrison and Evans 1987). Gametoclonal variation can be explained by a variety of processes, including new genetic variation brought about by cell culture techniques, segregation and independent assortment, chromosome doubling, and new variation brought about at the diploid level that result in heterozygosity (Morrison and Evans 1987; Huang 1992). Triploid plants produced through anther culture may be very profitable in crops where consumers prefer fruits without seeds. *Citrus clementina* Hort. ex Tan, *Carica papaya* L., and *Pyrus pyrifolia* Nakai have all been reported to produce triploids when regenerated from anther culture (Germana 2010).

2.6 Conclusions and Prospects

A very fruitful modern breeding system is haploid and doubled haploid breeding. A greater understanding of the various methods employed in double haploid production has been made possible by numerous studies. An extremely useful method for achieving extreme homozygosity and shortening the time needed to develop significant genotypes in many important plant species is anther culture. Techniques have been progressively developed and continuously improved since the discovery of haploid plants in 1920 and especially after the discovery of in vitro androgenesis in 1964. The approach has been standardized in several plant species and is being used in breeding programs. The majority of species for which there are established protocols are field crops or vegetables, but other plant species, such as ornamental and fruit plants, as well as other perennials, are also gradually being experimented to produce DHs. Nowadays, a wide range of breeding techniques are available to create more genetic variation and recover the desired gene combinations. Since thousands of pollen grains with distinct genotypes are carried by each anther of an inter-varietal/interspecific hybrid, these can quickly produce stable DH plantlets. In order to recover homozygous plants with uncommonly desirable gene combinations and speed up the breeding process in response to climate change, anther culture is a highly reproducible technique. The populations of DH derived from anther cultures might be receptive to molecular mapping of important genes and isolation of high-yielding, disease-resistant, and improved quality plant types.

References

Badiger BG (2017) Chapter 5: Basic principles and recent advances in anther/pollen culture for crop improvement. In: Plant Biotechnol, vol 1

Bagheri N, Jelodar NB (2008) Combining ability and heritability of callus induction and green plant regeneration in rice anther culture. Biotechnology 7(2):287–292

Bagni N, Tassoni A (2001) Biosynthesis, oxidation and conjugation of aliphatic polyamines in higher plants. Amino Acids 20:301–317

Barnabas B, Phaler PL, Kovacs G (1991) Direct effect of colchicines on the microspore embryogenesis to produce dihaploid plants in wheat (Triticum aestivum L.). Theor Appl Genet 81:675–678

Bhattacharya A, Mikkilineni V, Verma L, Palan B, Mali K, Char B (2014) Evaluation of doubled haploid culture conditions and regeneration of an *Indica* rice hybrid. Indian J Genet 74:384–386

Bhojwani SS, Razdan MK (1983) Plant tissue culture: theory and practice. Elsevier, New York, p 5

Binarova P, Straatman K, Hause B, Hause G, van Lammeren AAM (1993) Nuclear DNA synthesis during the induction of embryogenesis in cultured microspores and pollen of *Brassica napus* L. Theor Appl Genet 87:9–16

Bourgin JP, Nitsch JP (1967) Obtaining *Nicotiana* haploides from stamens cultivated *in vitro*. Ann Physiol Veg 9:377–382

Castillo AM, Cistué L, Vallés MP, Soriano M (2009) Chromosome doubling in monocots. In: Touraev A, Forster BP, Jain SM (eds) Advances in haploid production in higher plants. Springer, Dordrecht, pp 329–338. https://doi.org/10.1007/978-1-4020-8854-4_27

Chen CC (1977) In vitro development of plants from microspores of rice. In Vitro Cell Dev Biol Plant 13(8):484–489

Chopkar R, Jha Z, Verulkar S (2016) Induction of haploid plants in *indica* rice varieties through anther culture progressive research. Int J 11:596–599

Chu C (1978) The N6 medium and its applications to anther culture of cereal crops. In: Proceedings of symposium on plant tissue culture. Science Press, Peking, pp 43–50

Chu CC, Wang CC, Sun CS, Hsu C, Yin KC, Chu CY, Pi FY (1978) Establishment of an efficient medium for anther culture of rice through comparative experiments on the nitrogen sources. Sci Sin 18:659–668

Cistue L, Valles MP, Echavarri B, Sanz JM, Castillo AM (2003) Barley anther culture. In: Maluszynsky M, Kasha KJ, Forster BP, Szaejko I (eds) Doubled haploid production in crop plants. A manual. Kluwer/FAO-IAEA, Dordrecht/Vienna, pp 29–34

CRRI (2008) Annual Report 2007–2008: 34, National Rice Research Institute, Cuttack, India

CRRI (2010) Annual Report 2009-2010: 35, National Rice Research Institute, Cuttack, India.

Custers JBM, Cordewener JHG, Noellen Y, Dons HJM, van Lookeren Campagne MM (1994) Temperature controls both gametophytic and sporophytic development in microspore cultures of *Brassica napus*. Plant Cell Reprod 13:267–271

D'Amato F (1977) Cytogenetics of differentiation in tissue and cell cultures. In: Reinert J, Bajaj YPS (eds) Applied and fundamental aspects of plant cell, tissue, and organ culture. Springer, Berlin, pp 343–356

Dash B, Bhuyan SS, Singh SK, Chandravani M, Swain N, Rout P (2022) Androgenesis in indica rice: a comparative competency in development of doubled haploids. PLoS One 17(5):e0267442. https://doi.org/10.1371/journal.pone.0267442

Davies PA, Morton S (1998) A comparison of barley isolated microspore and anther culture and the influence of cell culture density. Plant Cell Rep 17(206):210
Dewi IS, Purwoko BS (2016) Role of polyamines in inhibition of ethylene biosynthesis and their effects on Rice anther culture development. Indones J Agric Sci 9(2):60–67
Dunwell JM (2010) Haploids in flowering plants: origins and exploitation. JPlant Biotechnol 8:377–424
El-Hennaway MA, Abdalla AF, Shafey SA, Al-Ashkar IM (2011) Production of doubled haploid wheat lines (Triticum aestivum L.) using anther culture technique. Ann Agric Sci 56(2):63–72
Evans DA, Sharp WR, Medina-Filho HP (1984) Somaclonal and gametoclonal variation. Am J Bot 71:759–774
Forster BP, Thomas WTB (2003) Doubled haploids in breeding. In: Maluszynski M, Kasha KJ, Forster BP, Szarejko I (eds) Doubled haploid production in crop plants: a manual. Kluwer Academic Publishers, Dordrecht, pp 337–350
Fukushima E, Iwasa S (1964) A spontaneous haploid plant in a certain *Brassico-Raphanus* hybrid strain. Proc Jap Acad 40:232–235
Gamborg OL, Miller RA, Ojima K (1968) Nutrient requirement suspension cultures of soybean root cells. Exp Cell Res 50:151–158
Germana MA (2010) Anther culture for haploid and doubled haploid production. Plant Cell Tissue Organ Cult 104:283–300
Greenblatt IM, Bock M (1967) A commercially desirable procedure for detection of monoploids in maize. J Hered 58:9–13
Grewal RK, Lulsdorf M, Croser J, Ochatt S, Vandenberg A, Warkentin TD (2009) Doubled-haploid production in chickpea (Cicer arietinum L.): role of stress treatments. Plant Cell Rep 28:12891299
Gueye T, Nadir KN (2010) *In vitro* production of double haploid plants from two rice species (*Oryza sativa* L. and *Oryza glaberrima* Steudt.) for the rapid development of new breeding material. Sci Res Essays 5(7):709–713
Guha S, Maheshwari SC (1964) In vitro production of embryos from anthers of Datura. Nature 204:497–498
Heberle-Bors E (1984) Genotypic control of pollen plant formation in *Nicotiana tabacum* L. Theor Appl Genet 68:475–479
Heberle-Bors E (1985) In vitro haploid formation from pollen: a critical review. Theor Appl Genet 71:361–374
Hofer M, Gomez A, Aguiriano E, Manzanera JA, Bueno MA (1999) Analysis of simple sequence repeat markers in homozygous lines in apple. Plant Breed 121:159–162
Huang B (1992) Genetic manipulation of microspores and microspore-derived embryos. In Vitro Cell Dev Biol 28:53–58
Jacquard C (2003) Microspore embryogenesis and programmed cell death in barley: effects of copper on albinism in recalcitrant cultivars. Plant Cell Rep 28:1329–1339
Jacquard C, Asakaviciute R, Hamalian AM, Sangwan RS, Devaux P, Clement C (2003) Barley anther culture: Effects of annual cycle and spike position on microspore embryogenesis and albinism. Plant Cell Rep 25:375–381
Kumar O, Choudhary M (2020) Double haploid: An overview. Int J Curr Microbiol App Sci 9(1):1012–1029
Kumlay AM, Ercisli S (2015) Callus induction, shoot proliferation and root regeneration of potato (Solanum tuberosum L.) stem node and leaf explants under long-day conditions. Biotechnol Biotechnol Equip 29(6):1075–1084
Luthar Z, Bohanec B (1999) Induction of direct somatic organogenesis in onion (*Allium cepa* L.) using a two-step flower or ovary culture. Plant Cell Rep 18:797–802
Magoon ML, Khanna KR (1963) Haploids. Caryologia 16:191–234
Maluszynski J (2003) Cytogenetic tests for ploidy level analyses -chromosome counting. In: Maluszynski M, Kasha KJ, Forster BP, Szarejko I (eds) Doubled haploid production in crop plants: A manual. Kluwer Academic Publishers, Dordrecht, pp 391–395.

Maluszynski M, Kasha KJ, Forster BP, Szarejko I (eds) (2003) Doubled haploid production in crop plants: a manual. Kluwer Academic Publishers. ISBN:1-4020-1544-5

Mayakaduwa R, Silva T (2023) Haploid induction in Indica rice: exploring new opportunities. Plan Theory 12(17):3118. https://doi.org/10.3390/plants12173118

Minj SS, Lee S, Park H, Park GH, Kim M (2016) Comparison of the efficiency to one step and two-step anther culture in rice (*Oryza sativa* L). J Environ Agric Sci 4(1):77–82

Mishra R, Rao RN, Rao GJN (2011) Anther culture response of Indica rice hybrids. Oryza 48:375–377

Mohiuddin AKM, Chowdhury MKU, Zaliha CA, Suhaimi N (2014) Influence of silver nitrate (ethylene inhibitor) on cucumber in vitro shoot regeneration. Plant Cell Tissue Organ Cult 51:75–78

Morinaga T, Fukushima E (1931) Preliminary report on the haploid plant of rice, *Oryza sativa* L. Proc Imp Acad 7:383–384

Morrison RA, Evans DA (1987) Gametoclonal variation. Plant Breed Rev 5:359–391

Murashige T, Skoog S (1962) A revised medium for rapid growth and bioassays with tobacco tissue cultures. Physiol Plant 15:473–497

Naik N, Rout P, Verma RL, Ngangkham U, Katara JL, Sahoo K, Singh ON, Samantaray S (2017) Development of promising doubled haploids from Indica rice hybrid: evaluation of uniformity and stability through morphological and molecular markers. Int J Agric Sci 8(47):1987–1992

Nishi T, Mitsuoka S (1969) Occurrence of various ploidy plants from anther and culture of rice plant. Jpn J Genet 44:341–346

Nitsch C, Nitsch B (1969) Effect of heat shock on the embryogenic power of *Datura innoxia* pollen grown in the anther or isolated from the anther. C R Acad Sci Paris 276D:303–306

Ochatt SJ (2008) Flow cytometry in plant breeding. Cytom Part A 73:581–598

Patel PN (2017) Rice doubled haploid production and homozygosity testing using phenotypic and molecular approaches. MSc thesis, IGKV, Raipur

Pelletier PJ, Caventou JB (1820) Note on the *Adipocirous* nature of ambergris and on the origin of this product. J Pharm 6:49–58

Powell W (1990) Environmental and genetic aspects of pollen embryogenesis. In: Bajaj YPS (ed) Biotechnology in agriculture and forestry. Part I. Haploids in crop improvement, vol 12. Springer, Berlin, pp 44–65

Przywara L, Pandey KK, Sanders PM (1988) Length of stomata as an indicator of ploidy level in *Actinidia deliciosa*. Newland J Bot 26:179–182

Pusadkar P, Jha Z (2018) In vitro colchicine treatment to increase the plant regeneration from *Indica* rice anther culture. J Pharmacogn Phytochem 7(4):440–443

Raina SK, Zapata FJ (1997) Enhanced anther culture efficiency of indica rice (*Oryza sativa* L) through modification of the culture media. Plant Breed 116:305–315

Ramiah K, Parthasarathi H, Ramanujam S (1935) Polyembryonie in rice (Oryza sativa). Ind J Agric Sci 5:119–124

Rao PS, Suprasanna P (1996) Methods to double haploid chromosome numbers. In Jain SM, Sopory SK, Veilleux RE (eds) In vitro haploid production in higher plants, vol 1. Kluwer Academic Publisher, Dordrecht, pp 317–339.

Raquin C (1983) Utilization of different sugars of carbon sources for *in vitro* anther culture of Petunia. Z Pflanzenzucht 111:453–457

Rout P, Naik N, Ngangkham U, Verma RL, Katara JL, Singh ON, Samantaray S (2016) Doubled haploids generated through anther culture from an elite long duration rice hybrid, CRHR32: method optimization and molecular characterization. Plant Biotechnol J 33:177–186

Rukmini M, Rao GJN, Rao RN (2011) Effect of cold pretreatment and phytohormones on anther culture efficiency of two indica rice (*Oryza sativa* L.) hybrids-Ajay and Rajlaxmi. Exp Biol Agric Sci 1:69–76

Sahoo SA, Jha Z, Verulkar SB, Srivastava AK, Suprasanna P (2019) High-throughput cell analysis based protocol for ploidy determination in anther-derived rice callus. Plant Cell Tissue Organ Cult 137:187–192

Shariatpanahi ME, Bal U, Heberle-Bors E, Touraev A (2006) Stresses applied for the reprogramming of plant microspores towards *in vitro* embryogenesis. Physiol Plantarum 127:519–534
Shrestha S, Kang WH (2016) Stomata length and density as an indicator of ploidy level in sweet pepper (*Capsicum annuum* L.). Res J Recent Sci 5(ISC-2015):4–10
Silva TD, Ratnayake WJ (2009) Anther culture potential of *indica* rice varieties, Kurulu Thuda and BG250. Trop Agric Res Ext 12(2):53–56
Sood S, Kumari V (2019) Efficient callus induction through anther culture in cultivars of *Brassica campestris var. Brown Sarson*. Int J Curr Microbiol Appl Sci 8(5):1003–1012
Sriskandarajah S, Sameri M, Lerceteau-Kohler E, Westerbergh A (2015) Increased recovery of green doubled haploid plants from barley anther culture. Crop Sci 55(6):2806
Sunderland N (1971) Anther culture: a progress report. Sci Prog (Oxford) 59:527–549
Tekam Y (2017) In vitro induction of chromosome doubling for the production of doubled haploid rice and homozygosity testing of morphological characterization and molecular approach. MSc thesis, IGKV, Raipur
Tesfaye M, Feyissa T, Gugsa L (2010) Embryogenic callus induction and regeneration in anther culture of noug (*Guizotia abyssinica* L.F Cass) Sinet. Ethiop J Sci 33(1):49–58
Thomas E, Hoffmann F, Wenzel G (1975) Haploid plantlets from microspores of rye. Z P flanzenzucht 75(2):106–113
Touraev A, Ilham A, Vicente O, Heberle-Bors E (1996) Stress induced microspore embryogenesis from tobacco microspores: an optimized system for molecular studies. Plant Cell Rep 15:561–565
Vagera J, Novotny J, Ohnoutkova L (2004) Induced androgenesis in vitro in mutated populations of barley, *Hordeum vulgare*. Plant Cell Tissue Org 77:55–61
Vasil IK (1973) The new biology of pollen. Naturwissenschaften 60:247–253
Weber S, Unker F, Friedt W (2004) Improved doubled haploid production protocol for Brassica napus using microspore colchicine treatment in vitro and ploidy determination by flow cytometry. Plant Breed 124:511–513
Wedzony M, Forster BP, Zur I, Golemiec E, Szechynska Hebda M, Dubas E, Gotebiowska, G (2009) Progress in doubled haploid technology in higher plants. In: Touraev A, Forster BP, Jain SM (eds) Advances in haploid production in higher plants. Springer, Berlin/Heidelberg, pp 1–34
Wenzel G, Foroughi-Wehr B (1984) Anther culture of cereals and grasses. In: Vasil IK (ed) Cell culture and somatic cell genetics of plants I. Academic, New York, pp 311–327
Zapata-Arias FJ (1997) Laboratory protocol for anther culture technique in rice. In: Maluszynski M, Kasha KJ, Forster BP, Szarejk I (eds) Doubled haploid production in crop plants, a manual. Kluwer Academic Publishers, Dordrecht, pp 109–116
Zhang YX, Chevreau E, Lespinasse Y (1989) Resulting from parthenogenes induced in situ by irradiated pollen and in vitro cultivation of immature pine trees. Obtaining haploid plants of apple (Malus X domestica Borkh.). CR Acad Sci Paris 307:451–457
Zhao JP, Simmonds DH, Newcomb W (1996) Induction of embryogenesis with colchicine instead of heat in microspores of *Brassica napus* L. cv. *Topas*. Planta 198:433–439
Zhu DY, Pan XG (1990) Rice (*Oryza sativa* L.) Guan 18: An improved variety through anther culture. In: Bajaj YPS (ed) Biotechnology in agriculture and forestry rice, vol 12. Springer, Berlin/Heidelberg, pp 04–211
Zlesak DC (2009) Pollen diameter and guard cell length as predictors of ploidy in diverse rose cultivars, species, and breeding lines Zlesak D.C. Roses. Floricult Ornam Biotechnol 3(1):53–70

Induction of Maternal Haploids in Plants

3

Tanveen Kaur, Vishal Sharma, and Suprasanna Penna

Abstract

Maternal haploids, harboring only a single set of chromosomes from the mother plant, have revolutionized plant breeding and crop improvement. By circumventing the lengthy process of natural chromosome doubling, these maternal haploids dramatically accelerate the breeding cycle, resulting in the development of completely homozygous doubled haploid (DH) lines. We explore how maternal haploids contribute to the creation of pure lines, robust hybrids, and novel varieties tailored to address environmental challenges. Their ability to efficiently transfer desirable traits, such as pest and disease resistance, drought tolerance, and enhanced yield potential, makes them invaluable tools in breeding. Integration of maternal haploids into hybrid seed production streamlines the process and reduces costs, further enhancing their economic impact. Beyond practical applications, the scientific contributions of maternal haploids cover elucidation of genomic imprinting phenomena, ethical considerations surrounding their use, and their potential for future integration with emerging technologies in maize breeding. This chapter comprehensively provides a critical overview of maternal haploids, highlighting their multifaceted potentials for the advancement of crop breeding, taking maize as an example, and offering a great scope for crop improvement.

T. Kaur
Roy J. Carver Department of Biochemistry, Biophysics and Molecular Biology, Iowa State University of Science and Technology, Iowa, USA

V. Sharma
Biotechnology Division, CSIR-Institute of Himalayan Bioresource Technology, Palampur, Himachal Pradesh, India

S. Penna (✉)
Amity Center for Nuclear Biotechnology, Amity Institute of Biotechnology, Amity University Maharashtra, Mumbai, India

Z. Jha et al. (eds.), *Doubled Haploids: Technological Advances and Role In Crop Improvement*, https://doi.org/10.1007/978-981-96-2339-6_3

Keywords

Maternal haploids · Hybrids · Pest and disease resistance · Yield potential · Genome editing · Hybrid seed production

3.1 Introduction

Maternal haploids contain only the maternal set of chromosomes. In other words, these are haploid plants that are derived from the egg cell of the female parent. Maternal haploids are of great interest to plant breeders and geneticists because they can be used to create homozygous diploid lines in a single generation. This is because the haploid state allows for the expression of recessive alleles, which can be masked in the diploid state. Maternal haploids are generally derived from the egg cell of the female parent. In *Zea mays*, maternal haploids are induced through in vivo induction techniques that use special genotypes known as haploid inducers (Liu et al. 2018). The haploid inducers have the characteristic to induce haploids. The effect of genotype on maternal haploid plant production in maize was studied. The frequency of gynogenetic plants when "Stock 6" is used as pollinator varies according to the female parent genotype. No simple relation was observed between genotypic aptitudes for gynogenetic and androgenetic development, which occurred after pollination of "W23" plant carrying the "indeterminate gametophyte" gene. Furthermore, the population NS, a favorably responsive genotype to anther culture, does not exhibit exceptional ability for in vivo gynogenesis. The effect of inbreeding and the influence of maternal haploid origin suggest that specific genes control maternal haploid initiation and development. However, gynogenetic development is not limited to a particular genotype. The frequency of maternal haploids may be increased by using specific pollen parents. Attempts were made to select for a high haploidy-inducing trait, and the present study reports the successful development of lines that can be utilized as pollen parents to induce haploids for experimental purposes and breeding programs. When an inbred line "WS14," derived from the cross W23 × Stock 6, was used as pollen parent, 2–5% maternal haploids were obtained according to the female parent genotype. A high haploidy-inducing potential is a heritable trait and may be controlled by a limited number of genes. Genetic determination of the haploidy-inducing character was examined in relation to the efficiency of the selecting method and the mechanisms involved in the origin of maternal haploids.

This chapter with a specific focus on maize (*Zea mays*) endeavors to provide a comprehensive understanding of various facets within this field of study. By delving into the intricacies of female gametophyte development in maize, the chapter aims to unravel molecular and genetic factors influencing haploid embryo formation. Additionally, it explores the dynamics of genomic imprinting in maize, shedding light on its broader implications in comparable processes across different crops and animals. The investigation extends to uncovering the genetic basis and ecological context of maternal haploidy in maize and its wild relatives, offering insights into the adaptive strategies of this essential crop. Utilizing maternal haploids in maize becomes a central theme, with a focus on identifying novel traits not expressed in the diploid state, particularly emphasizing recessive traits that gain dominance in the haploid state. Furthermore, the historical significance of maternal haploid induction in maize is underscored as an indispensable tool in maize breeding and genetics research. Addressing challenges and limitations in maize cultivation, the chapter ultimately seeks to contribute to global maize production enhancement through a nuanced comprehension and application of maternal haploids, thereby advancing crop improvement and hybrid seed production.

3.1.1 Historical Perspective

The induction of maternal haploids in plants has been a subject of research for over a century. The first successful induction of haploids in maize was reported by Blakeslee and Avery in 1937 (Lashermes and Beckert 1988). Since then, the technique has been refined and improved, and it has become an essential tool in maize breeding and genetics research. The historical significance of maternal haploid induction in maize lies in its ability to produce haploid plants from the female gametophyte, which is otherwise difficult to achieve. The technique has been used to study the genetic control of haploid induction, to develop haploid inducer lines, and to produce doubled haploid lines for use in breeding programs (Lashermes and Beckert 1988; Haixiao et al. 2016).

The development of haploid inducer lines has been a significant outcome of research on maternal haploid induction in maize. These lines are used to induce haploids in maize by pollinating the female parent with pollen from the inducer line. The frequency of haploid induction varies depending on the genotype of the female parent, but it can be increased by using specific pollen parents (Lashermes and Beckert 1988). The efficiency of haploid induction is influenced by several factors, including the genotype of the female parent, the genotype of the inducer line, and the environmental conditions (Haixiao et al. 2016). The genetic control of maternal haploid induction in maize has been studied extensively. Specific genes have been identified that control maternal haploid initiation and development (Lashermes and Beckert 1988). The haploidy-inducing character is a heritable trait that may be controlled by a limited number of genes (Lashermes and Beckert 1988; Wu et al. 2014). The mechanisms involved in the origin of maternal haploids are still not fully understood, but it is believed that specific genes control maternal haploid initiation and development (Lashermes and Beckert 1988).

One example of the use of maternal haploid induction in maize research is the study of the effect of genotype on maternal haploid plant production. The frequency of gynogenetic plants varies according to the female parent genotype, and no simple relation has been observed between genotypic aptitudes for gynogenetic and androgenetic development (Lashermes and Beckert 1988). The effect of inbreeding and the influence of maternal haploid origin suggest that specific genes control maternal haploid initiation and development. Attempts have been made to select for a high haploidy-inducing trait, and lines have been developed that can be utilized as pollen parents to induce haploids for experimental purposes and breeding programs (Lashermes and Beckert 1988). Another example of the use of maternal haploid induction in maize research is the study of the genetic basis of haploid induction. The molecular mechanisms underlying haploid induction in maize are still unclear, but research has identified several genes that are involved in the process. The efficiency of haploid induction is influenced by several factors, including the genotype of the female parent, the genotype of the inducer line, and the environmental conditions (Haixiao et al. 2016). The development of haploid inducer lines has been a significant outcome of research on maternal haploid induction in maize (Lashermes and Beckert 1988).

The maternal haploid induction in maize has a rich history and has been an essential tool in maize breeding and genetics research. The technique has been used to study the genetic control of haploid induction, to develop haploid inducer lines, and to produce doubled haploid lines for use in breeding programs. The genetic control of maternal haploid induction in maize has been extensively studied, and specific genes have been identified that control maternal haploid initiation and development. The efficiency of haploid induction is influenced by several factors, including the genotype of the female parent, the genotype of the inducer line, and the environmental conditions. The development of haploid inducer lines has been a significant outcome of significant research efforts on maternal haploid induction in maize. Furthermore, a detailed history of maternal haploid production in *Zea mays* plants has been detailed in Table 3.1.

3.2 Haploid Formation Mechanisms

Maize, a staple crop with significant agricultural importance, exhibits various mechanisms for haploid formation, contributing to the advancement of maize breeding techniques. In this section, we delve into the key mechanisms involved in inducing haploids in maize, focusing on notable approaches such as Stock 6, CENH3-mediated haploid induction, in vitro haploid induction, in planta haploid induction, and interspecies hybridization.

3.2.1 Stock 6 Mechanism

Coe's study using the haploid inducer "Stock 6" represents the main advancement in haploid breeding of maize. Stock 6 is a haploid inducer that produces haploids in maize. It is a maternal haploid inducer that is used to produce haploids in maize (Indu et al. 2023). The haploid formation mechanism involves hetero-fertilization together with failed egg-sperm cell fusion. The genes involved in this mechanism are still unclear (Tian et al. 2018).

Table 3.1 Historical perspective of haploid induction in maize

Year	Scientists involved	Events	Why	How	Outcome	References
1929	L. F. Randolph and L. J. Stadler	First report of haploid plants in maize	To study the genetic basis of maize	–	Discovered the phenomenon of haploidy in maize	Trentin et al. (2020)
1947	Chase	Natural occurrence of haploid plants in commercial inbred lines	To develop homozygous diploids of maize from monoploids	Pollination with Stock 6, using the R1-nj allele to identify haploids	Reported haploid frequencies lower than 0.1%	Chase (1947)
1952	Chase	Production of homozygous diploids of maize	To develop a method for producing homozygous diploids that would be faster and more efficient than traditional breeding methods	Isolating monoploids in quantity, growing the plants, and deriving homozygous diploids from them through selfing	Development of a method for producing homozygous diploids that was faster and more efficient than traditional breeding methods	Chase (1952)
1959	Coe	A line of maize with high haploid frequency (3.2%)	To facilitate the production of haploids for experimental purposes	Selection from a cross of W23 × Stock 6	Achieved haploid frequencies of 3.2%	Coe (1959)
1969	Kermicle	Indeterminate gametophyte1 (ig1) gene identified as the cause of paternal haploid induction	To explain the origin of paternal haploids in maize	By crossing a line carrying the ig1 gene with normal lines	Found that the ig1 gene causes the development of haploid embryos from unfertilized egg cells	Kermicle (1969)
1978	Aman and Sarkar	Selection for haploidy-inducing potential in maize	To improve the efficiency of haploid induction	Selection from a cross of W23 × Stock 6	Obtained haploid frequencies of 5.5%. Developed a line with 9.6% HIR	Aman and Sarkar (1978)

(continued)

Table 3.1 (continued)

Year	Scientists involved	Events	Why	How	Outcome	References
1988	Lashermes and Beckert	Genetic control of maternal haploidy in maize and selection of haploid inducing lines	To understand the genetic factors affecting haploid induction	Pollination with WS14, a derivative of W23 × Stock 6	Identified specific genes controlling maternal haploid initiation and development	Lashermes and Beckert (1988)
1997	Deimling et al.	Genetic analysis of maternal haploid induction in maize	To map quantitative trait loci (QTLs) for haploid induction rate	Pollination with QTL-NILs derived from a cross of W23 × Stock 6	Detected three QTLs on chromosomes 1, 5, and 10	Deimling et al. (1997)
1999	Röber	QTL mapping of maternal haploid induction in maize	To identify quantitative trait loci which contribute to the maternal genetics of haploid induction	–	Two quantitative trait loci (QTLs) were identified, qmhir1 (between the flanking marker loci umc1292 and bnlg1014, and it explained 14.70% phenotypic variation) and qmhir2 (between marker loci umc1844 and umc2277 and it explained 8.42% of the phenotypic variation). Genetic effect of both is partial dominance	Röber (1999)
2010	Rotarenco et al.	A line of maize with very high haploid frequency (14.5%)	To increase the haploid induction rate and reduce the cost of doubled haploid production	Selection from a cross of W23 × Stock 6	Achieved haploid frequencies of 14.5%	Rotarenco et al. (2010a, b)

2012	Prigge et al.	MATRILINEAL gene identified as a major factor for maternal haploid induction in maize	To elucidate the molecular mechanism of haploid induction	Fine-mapping and candidate gene analysis	Discovered a mutation in a phospholipase gene that disrupts the fertilization process	Prigge et al. (2012)
2012	Prigge et al.	ZmDMP gene identified as another major factor for maternal haploid induction in maize	To identify the ZmDMP gene as another major factor for maternal haploid induction in maize	performing a SNP-based genome-wide association study using a large and diverse panel of 159 maternal haploid inducers	Identification of a major gene near MTL, which is present in all inducers and necessary to disrupt haploid induction; single-nucleotide change in ZmDMP leads to a two- to threefold increase in the haploid induction rate	Prigge et al. (2012)
2015	Liu et al.	Genome-wide association study of maternal haploid induction in maize	To further explore the genetic basis of haploid induction	Fine-mapping and candidate gene analysis	Discovered a mutation in a DNA methyltransferase gene that affects the endosperm development	Liu et al. (2015)
2017	Gilles et al.	Kokopelli gene identified as a factor for maternal haploid induction in *Arabidopsis*	Study hypothesized that other mutants impaired in gamete interactions might also induce haploid embryos, similarly to what had been reported for the Atdmp8/9 mutant	–	Search for genes that were expressed in the *Arabidopsis* male gametophyte and known to lead to single fertilization events when mutated identified AtKPL (KOKOPELLI, AT5G63720) as a candidate	Gilles et al. (2017)
2023	Trentin et al.	Genetic basis of maize maternal haploid induction beyond MATRILINEAL and ZmDMP	To identify additional genes or QTL for haploid induction	Genome-wide association study using a panel of 159 maternal haploid inducers	Identified a QTL on chromosome 10 harboring a kokopelli orthologue and several QTLs with smaller effects	Trentin et al. (2023)

3.2.2 CENH3-Mediated Haploid Induction

The utilization of CENH3, a centromere-specific histone H3 variant, represents an innovative approach to induce haploid formation in maize. This mechanism involves the development of a haploid-inducing transgenic line expressing CENH3 under the control of a pollen-specific promoter. Crossing this transgenic line with a non-transgenic line results in haploid progeny. The genes implicated in this process include CENH3 and other regulatory genes associated with meiosis (Meng et al. 2021). For instance, in a study conducted by Liu et al. (2018), the effectiveness and stability of the CENH3-mediated haploid induction approach in generating haploid maize plants were demonstrated. The researchers employed a haploid-inducing transgenic line expressing CENH3 regulated by a pollen-specific promoter, which was subsequently crossed with a non-transgenic line to yield haploid progeny with notable efficiency and stability (Rai et al. 2023). Similarly, findings from a separate study by Li et al. (2021) supported the utility of the CENH3-mediated haploid induction method for developing haploid maize plants characterized by high frequency and purity. Employing a haploid-inducing transgenic line expressing CENH3 under the control of a pollen-specific promoter, the researchers executed crosses with non-transgenic lines to generate haploid progeny. Furthermore, the haploid plants obtained were subjected to doubling, resulting in the production of homozygous diploid plants (Wang et al. 2021). These studies collectively underscore the potential of the CENH3-mediated haploid induction strategy as a valuable tool for the efficient and stable development of haploid maize plants, offering insights into advancements in plant breeding methodologies.

3.2.3 In Vitro Haploid Induction

In vitro culture techniques offer another avenue for inducing haploid formation in maize (Genovesi 1990). The mechanism entails the cultivation of immature embryos or anthers in a nutrient medium containing plant growth regulators (Guha and Maheshwari 1964). Subsequent regeneration of haploid embryos into plants is achieved through this in vitro approach. While the specific genes responsible for in vitro haploid induction in maize are not yet elucidated, ongoing research aims to unravel the genetic components underlying this method. For instance, in the pioneering work of Ao et al. (1982), unpollinated maize ovaries were cultured on MS and N6 medium, resulting in the successful generation of nearly 2% haploid plants. The primary objective of this study was to establish a protocol for the production of haploid plants suitable for integration into breeding programs. Building upon this foundation, Barloy (1989) conducted research comparing the aptitude for anther culture in specific androgenetic doubled haploid maize lines (Büter 1997). The overarching goal was to refine a protocol for generating haploid plants applicable to breeding initiatives. Subsequently, Petolino (1992) achieved the production of haploids through tissue culture from megasporocytes and microsporocytes of maize, emphasizing the continual effort to develop a robust protocol for generating haploid plants in the context of breeding programs. In all three instances, researchers employed in vitro culture techniques to induce haploid formation in maize. This involved cultivating immature embryos or anthers in a nutrient medium enriched with plant growth regulators. The subsequent regeneration of haploid embryos into viable plants was successfully accomplished through this in vitro approach. Although the specific genes responsible for in vitro haploid induction in maize remain undisclosed, ongoing research endeavors aim to unravel the genetic components that underlie this method, contributing to the advancement of maize breeding programs (Meng et al. 2021; Watts et al. 2018).

3.2.4 In Planta Haploid Induction

In planta techniques are employed to induce haploid formation in maize, involving the pollination of maize plants with haploid inducers. Haploid seeds are identified using a seed coloration marker system, and haploid embryos are then doubled by treating seedlings with a mitotic inhibitor. Despite the effectiveness of this approach, the genes associated with in planta haploid induction in maize remain elusive and necessitate further investigation. Researchers at both the University of Wisconsin-Madison and the University of Illinois have independently pursued the development of more efficient methods for producing haploid maize plants using in planta haploid induction techniques. At the University of Wisconsin-Madison, scientists focused on the haploid inducer line, Stock 6. The protocol involved pollinating maize plants with Stock 6 and identifying haploid seeds through a seed coloration marker system. Subsequently, haploid embryos were doubled by treating seedlings with a mitotic inhibitor. The specific genes associated with in planta haploid induction in maize were not identified, indicating the need for further investigation in this aspect. The University of Illinois researchers employed a similar approach, utilizing the haploid inducer line, RWS, for inducing haploid formation in maize. The procedure included pollination with RWS, seed identification via a coloration marker system, and doubling of haploid embryos with a mitotic inhibitor. Similar to the Wisconsin study, the genes related to in planta haploid induction in maize were not uncovered, emphasizing the ongoing challenges in this area. Despite the elusive genetic information, both studies successfully demonstrated the effectiveness of their in planta haploid induction techniques, showcasing the potential advantages of these approaches in efficiently producing haploid maize plants compared to other methods (Meng et al. 2021; Trentin et al. 2020, 2023; Jacquier et al. 2021). The collective findings underscore the importance of continued research to unravel the genetic intricacies of in planta haploid induction in maize.

3.2.5 Interspecies Hybridization

Interspecies hybridization serves as an intriguing method for inducing haploid formation in maize. This mechanism involves crossing maize with a related species capable of producing haploid embryos. Subsequent doubling of haploid embryos is achieved through the application of a mitotic inhibitor. However, the specific genes responsible for interspecies hybridization-induced haploid formation in maize are yet to be fully understood. In conclusion, the diverse mechanisms discussed highlight the complexity and richness of haploid formation in maize. While significant progress has been made, ongoing research endeavors are essential to elucidate the underlying genetic components and further refine these techniques for the improvement of maize breeding programs.

3.3 Gametophytic Pathways

Maize is an important crop that is widely cultivated around the world. The production of haploids via gametophytic pathways is an interesting biological phenomenon in maize that has the potential to greatly accelerate maize breeding. In this section, we will discuss the mechanisms of gametophytic haploid production in maize, with a focus on maternal haploid formation.

3.3.1 Overview of Gametophytic Haploid Production

The unique ability to generate haploid plants, cells containing only one set of chromosomes, offers a powerful tool for maize breeding and scientific exploration. While several methods exist, gametophytic haploid production, particularly through maternal haploid induction via a male inducer, holds an immense potential due to its efficiency and intriguing biological phenomena. This section delves into the complexities of gametophytic haploid production in maize, exploring the mechanisms, challenges, and exciting possibilities it presents. At the heart of gametophytic haploid production in maize lies the enigmatic “haploid inducer” genotype. When crossed with a normal maize plant, these inducers lead to the formation of both diploid and haploid embryos on the same ear. The exact mechanism behind this “double fertilization” remains shrouded in mystery, though both male gametophytic and zygotic selection are believed to play crucial roles. In a study by Xu et al. (2013), a locus named segregation distortion 1 (sed1) was identified as a key player in this process. Interestingly, sed1 expression was localized to the male gametophyte, highlighting its involvement in influencing pollen behavior and potentially disrupting double fertilization. This discovery not only sheds light on the haploid induction process but also unveils a fascinating example of segregation distortion, a powerful evolutionary force capable of altering allele frequencies within populations. Doubled haploids, plants regenerated from haploid cells through chromosome doubling, provide an immense value for maize breeding. By eliminating genetic masking through homozygosity, doubled haploid lines allow for rapid selection of desirable traits and facilitate the acceleration of breeding cycles. This technology significantly shortens the time required to develop pure lines compared to traditional methods, thereby contributing to faster genetic gain and improved agronomic performance. Currently, various strategies, including in vitro and in planta techniques, are employed to generate haploids, followed by natural or chemical doubling to produce fertile doubled haploids. Recent breakthroughs in understanding the genetic and molecular basis of doubled haploidy have paved the way for even more precise and efficient breeding strategies (Chumakov and Mazilov 2022).

The ability to produce maize haploids extends beyond just accelerating breeding. Haploid lines serve as invaluable tools for genetic studies, enabling researchers to map genes, identify quantitative trait loci (QTLs), and unravel the complex interplay of genes and environment. Their inherently single-cell nature facilitates high-throughput genome sequencing and expression analysis, offering unparalleled insights into gene function and regulatory networks. Furthermore, doubled haploids play a crucial role in reverse breeding, allowing researchers to convert the existing hybrid lines into completely homozygous lines while maintaining their desirable traits. Additionally, they serve as powerful tools for exploring untapped germplasm and landraces, unlocking hidden genetic diversity for potential agronomic improvements. Despite its immense potential, gametophytic haploid production in maize faces certain challenges. The efficiency of haploid induction remains variable, often dependent on environmental factors and the specific inducer genotype employed. Segregation distortion associated with certain inducers further complicates breeding efforts. Addressing these challenges requires continued research and development of novel inducer lines with higher haploid induction rates and minimal distortion effects. Additionally, exploring alternative haploid production methods, such as those utilizing gametocyte cultures or meiotic mutants, holds a promise for further refinement and optimization of the technology. The future of gametophytic haploid production in maize appears bright. With ongoing research into the mechanisms underlying haploid induction and doubled haploidy, alongside the development of new technologies and marker-assisted techniques, this powerful tool is poised to revolutionize maize breeding and scientific exploration. By delving deeper into the intricacies of gametophytic haploid production, we can unlock a hidden world of genetic potential and pave the way for a new era of maize improvement and scientific discovery.

3.3.2 Types of Gametophytic Haploid Formation

The induction of maternal haploids in maize (*Zea mays* L.) offers a powerful tool for plant breeding and genetic research. Haploid plants, containing only a single set of chromosomes, can be readily doubled to produce completely homozygous diploid lines, invaluable for trait mapping, mutant analysis, and hybrid development. While numerous methods exist for inducing haploidy in maize, they all ultimately rely on the formation of haploid gametes during microgametogenesis or megagametogenesis. This section delves into the different types of gametophytic haploid formation observed in maize. Following are the cells involved in the process of gametophytic haploid formation:

3.3.2.1 Antipodal Cells

In contrast to their ephemeral fate in many plants, antipodal cells in certain maize varieties exhibit a remarkable resilience. Not only do they persist after fertilization, but they can undergo extensive division, reaching up to 100 cells with multiple nuclei (Chettoor and Evans 2015). This persistence is further underscored by their specialized cell walls, featuring invaginations that likely facilitate nutrient transfer to the developing embryo (Song et al. 2014). These unique characteristics suggest a potential role for antipodal cells beyond their initial function in guiding pollen tube growth. This notion is further supported by their sensitivity to mutations in specific genes, such as ARGONAUTE9, ZmDCL1, and ZmRDR2, which impact their development and function. Interestingly, antipodal cells also express genes associated with stress response, cell cycle regulation, and DNA repair, hinting at a complex interplay between their longevity and internal processes. However, their fate ultimately converges with that of most plant cells, undergoing programmed cell death at the late stages of embryo sac development (Dawit et al. 2017). This intriguing dichotomy—persistence and programmed death—underscores the multifaceted nature of these specialized cells and their potential contributions to seed development in specific plant lineages.

3.3.2.2 Egg Cells

Egg cells, the female gametes of plants, play a pivotal role in sexual reproduction. These specialized cells, surrounded by two synergids for pollen tube guidance, are derived from the micropylar cell of the four-nucleate embryo sac. Beyond their critical role in fertilization, recent research has revealed fascinating aspects of egg cell biology. Studies have identified genes expressed in egg cells that govern crucial processes like cell fate specification, signaling, and cell wall modification. Moreover, advancements in egg cell manipulation have paved the way for exciting possibilities. By tweaking CENH3 expression, researchers can induce egg cells to form haploid embryos, offering a novel approach for plant breeding and genetic studies (Trentin et al. 2023). Maize maternal haploid inducers are genotypes that have the inherent ability to generate seeds with haploid embryos when used to pollinate other genotypes (Gilles et al. 2020; Trentin et al. 2020). In maize, doubled haploid (DH) lines are created in vivo through crosses with maternal haploid inducers. Their induction ability, usually expressed as haploid induction rate (HIR), is known to be under polygenic control. Although two major genes (MTL and ZmDMP) affecting this trait were recently described, many others remain unknown. A recent study identified a major gene near MTL, which is present in all inducers and necessary to disrupt haploid induction. They also found a significant quantitative trait locus (QTL) on chromosome 10 using a case-control mapping approach, in which non-inducers were used as controls. These QTL could be incorporated into inducer breeding programs through marker-assisted selection (MAS) approaches. Further improving HIR is important to reduce the cost of DH line production (Trentin et al. 2023). Additionally, the ability to isolate and culture egg cells in vitro opens doors for further exploration of their developmental potential and utilization in biotechnological applications. In conclusion, egg cells stand as fascinating gatekeepers of plant reproduction, holding an immense potential for advancing our understanding and manipulation of plant development.

3.3.2.3 Synergids

Synergids, flanking the egg cell within the embryo sac, are crucial players in plant fertilization. These accessory cells orchestrate the arrival of the pollen tube through a multi-pronged approach. First, they secrete alluring molecules that guide the pollen tube toward the micropyle, the narrow opening into the ovule. Second, their filiform apparatus, a finger-like extension, interacts directly with the pollen tube, providing further navigational cues. Upon successful pollen tube entry, synergids undergo programmed cell death, paving the way for sperm delivery to the egg and central cell. This intricate process is underpinned by a dedicated gene expression program in synergids, regulating communication, differentiation, and ultimately their own demise. Moreover, synergids possess the remarkable ability to bypass the usual fertilization route, forming haploid embryos under the influence of specific male inducers, offering fascinating insights into their developmental potential.

3.3.3 Mechanisms of Gametophytic Haploid Formation

3.3.3.1 Diplosporous

In the realm of maize biology, the term "Diplosporous" refers to a specific reproductive mechanism that deviates from the conventional process of meiosis. This mechanism is characterized by the development of a haploid embryo directly from a diploid megaspore mother cell without undergoing the usual process of meiotic cell division. In simpler terms, diplosporous reproduction in maize involves the formation of a haploid embryo without the reduction of chromosome number typical in sexual reproduction. Interestingly, diplosporous haploids can emerge spontaneously, albeit at low frequencies, owing to errors that may occur during cell division or chromosome elimination processes. These sporadic occurrences contribute to the genetic diversity within maize populations. However, researchers have discovered that the frequency of diplosporous events can be significantly heightened through deliberate interventions. Specifically, by utilizing maize plants with specific genetic backgrounds or inducer lines carrying dominant genes capable of disrupting normal megaspore development, scientists can manipulate and increase the occurrence of diplosporous reproduction. This intentional induction provides a valuable tool for maize researchers and breeders, allowing them to explore novel genetic variations and potentially harness unique traits for crop improvement. Understanding diplosporous mechanisms in maize not only enriches our knowledge of plant reproduction but also opens avenues for targeted genetic modifications in agricultural practices.

3.3.3.2 Asporous

Apospory is a distinctive pathway in the context of maize reproduction, describing a process wherein a haploid embryo is formed directly from somatic cells of the nucellus, which is the nutritive tissue surrounding the megaspores. Unlike conventional reproductive processes involving meiosis and megagametogenesis, aposporous development circumvents these stages. The result is the creation of haploid embryos directly from the nucellar cells, without the need for fertilization. One notable characteristic of aposporous haploids in maize is the occurrence of twin seedlings. These seedlings consist of one diploid individual originating from the fertilized egg, adhering to the typical sexual reproduction pattern, and another haploid individual derived from the nucellar cell through aposporous development. Identification of aposporous haploids is essential for understanding maize reproduction, as it provides insights into alternative pathways that contribute to genetic diversity. However, it is noteworthy that the frequency of aposporous haploid formation is generally lower compared to diplosporous pathways, and it is often influenced by environmental factors, highlighting the complex interplay between genetic and environmental determinants in maize reproductive biology.

3.3.3.3 Twin Plants

In the context of maize, the phenomenon of twin plants represents a fascinating aspect of reproductive biology. This intriguing occurrence is characterized by the simultaneous development of two embryos within a single ovule. What sets twin plants apart is the unique combination of one diploid embryo and one haploid embryo coexisting within the same ovule. The haploid embryo, derived from various mechanisms such as diplospory, apospory, or the degeneration of a fertilized egg cell, introduces an additional layer of complexity to the maize reproductive process. The presence of twin plants holds significant implications for research and genetic analysis. The coexistence of diploid and haploid tissues within a single plant provides an invaluable resource for comparative studies. Researchers can explore the genetic and physiological variations between these two types of tissues, shedding light on the intricate mechanisms underlying maize development. Moreover, the study of twin plants facilitates a deeper understanding of the genetic factors influencing embryonic development, potentially contributing to advancements in crop breeding and improvement. In essence, twin plants in maize serve as a captivating subject for scientific inquiry, offering a dual perspective on genetic and developmental processes. The exploration of this phenomenon opens up new avenues for researchers to unravel the mysteries of maize reproduction, ultimately contributing to the enhancement of agricultural practices and the development of more resilient and productive maize varieties.

3.3.3.4 Androgenesis

Androgenesis, in the context of maize, is a noteworthy technique for haploid production that, although not strictly gametophytic, plays a crucial role in the development of haploid plantlets. This process involves the cultivation of isolated pollen grains under carefully controlled and specific conditions, prompting their transformation into haploid individuals. Unlike traditional methods, androgenesis boasts advantages in terms of simplicity and controllability. The technique allows for a more direct manipulation of the male reproductive cells, streamlining the process of generating haploid plantlets. However, it is essential to note that the success rate of androgenesis can vary significantly depending on the specific genotype of maize and the culture conditions employed. Despite this variability, the potential benefits in terms of efficiency and ease of manipulation make androgenesis a valuable tool in the broader scope of plant breeding and genetic research for maize.

3.3.3.5 Microspore-Derived Haploids

Microspore-derived haploids (Mdh) represent a significant aspect of gametophytic haploid formation in maize. The process begins with meiosis, wherein a diploid microspore mother cell (MMC) undergoes cell division, yielding four haploid microspores. Subsequently, these microspores can undergo one or two mitotic divisions, resulting in a pollen grain with two or three nuclei, respectively. This deviation from the standard pollen development pathway provides a unique opportunity for the generation of haploid embryos. Under stress conditions or with the application of specific chemicals, the haploid microspore or pollen grain can bypass its typical developmental program and initiate direct embryo development. This phenomenon is particularly valuable in maize breeding programs, as it allows for the creation of haploid plants with desirable traits in a more controlled and efficient manner. Microspore-derived haploids offer a powerful tool for researchers and breeders seeking to accelerate the development of improved maize varieties through the manipulation of the plant's reproductive processes.

3.3.3.6 Megaspore-Derived Haploids

Megaspore-derived haploids (Msh) represent a less common but intriguing pathway in maize reproduction, distinct from the more prevalent microspore-derived haploids (Mdh). The Msh pathway unfolds during megasporogenesis, the initial phase in female gamete development. Here, a diploid megaspore mother cell undergoes meiosis, generating four haploid megaspores. Remarkably, only one of these megaspores typically survives, proceeding to develop into the embryo sac, while the others undergo degeneration. Subsequently, during megagametogenesis, the surviving megaspore undergoes three mitotic divisions, culminating in the formation of an eight-nucleate embryo sac. The distinctive aspect of Msh lies in the subsequent embryo development, akin to Mdh. Under specific conditions, one of the nuclei within the embryo sac can deviate from its typical gametophytic developmental trajectory and instead directly initiate the formation of a haploid embryo. This intricate process adds a layer of complexity to maize reproduction, reflecting the diverse mechanism nature employs to ensure genetic variability and adaptability within the maize population.

3.3.3.7 Twin-Seed Haploids

In the realm of maize genetics, twin-seed haploids (Ts) represent a distinctive and intriguing phenomenon. This unique type of haploid formation involves the development of two haploid embryos from a single fertilized egg cell. The process leading to the creation of twin-seed haploids can manifest through various mechanisms, each contributing to the formation of distinct haploid embryos. One mechanism is mitotic nondisjunction, occurring during the initial mitotic division of the zygote. In this process, failure of chromosome separation results in two daughter cells with disparate chromosome numbers. Consequently, one of the daughter cells becomes haploid, while the other retains its diploid state. Anaphase bridge breakage is another mechanism leading to the formation of twin-seed haploids. During the anaphase stage of the first mitotic division, chromosomes can break and create a bridge between the two daughter cells. This can result in an uneven distribution of chromosomes, ultimately yielding one haploid daughter cell alongside another maintaining a diploid status. Furthermore, the loss of a chromosome during early embryo development represents yet another pathway to twin-seed haploid formation. In this scenario, the absence of a chromosome leads to the creation of a haploid embryo. The intricate processes underlying twin-seed haploid formation not only contribute to the diversity of maize genetics but also offer valuable insights into the mechanisms of embryonic development and chromosome behavior in this essential cereal crop. Understanding these mechanisms can potentially be harnessed for the improvement of maize breeding programs and the development of novel varieties with enhanced traits and productivity.

3.3.3.8 Parthenogenesis

Parthenogenesis is a fascinating phenomenon that involves the development of an embryo from an unfertilized egg cell. Although not strictly a gametophytic process, parthenogenesis in maize can lead to the formation of haploid embryos under certain conditions. This process, known as diplosporous parthenogenesis, involves the spontaneous doubling of chromosomes in the egg cell before embryonic development takes place. While not as prevalent as the more conventional method of obtaining haploid embryos through microspore-derived (MSD) technology, parthenogenesis can still occur in maize. Haploid embryos resulting from parthenogenesis may arise spontaneously, but they can also be induced by stress treatments or specific genetic mutations. This alternative pathway can be particularly advantageous in maize breeding programs, especially when dealing with recalcitrant genotypes that show resistance to MSD. The ability to generate haploid lines through parthenogenesis offers an alternative and potentially valuable approach for overcoming challenges in maize breeding and improving the efficiency of genetic manipulation in this important crop.

3.3.3.9 Microspore Elimination

Microspore elimination is a prominent method employed in maize breeding to induce haploidy, a crucial step in the development of doubled haploid lines. In MSD genotypes, the induction of haploidy is orchestrated by a recessive gene that disrupts the typical progression of male gametophyte development. This disruption specifically targets microspores, resulting in their degeneration either during meiosis or early stages of pollen development. Consequently, this process selectively eliminates microspores, leaving behind functional haploid megaspores. The remaining megaspores serve as the foundation for the development of haploid embryos upon fertilization with normal pollen. There exist two primary categories of MSD genes that play pivotal roles in this process. Sporophytic MSD genes, such as ms1 and ms4, are expressed in the sporophytic tissues surrounding the developing microspores. This expression leads to the degeneration of microspores and ensures the preferential survival of haploid megaspores. On the other hand, gametophytic MSD genes, including tsf and nuv, are expressed directly within the microspores themselves. The activation of these genes triggers programmed cell death within the microspores, contributing to the selective elimination of non-functional male gametophytes. Through the precise manipulation of these MSD genes, maize breeders can efficiently generate haploid lines, facilitating the development of homozygous lines and accelerating the breeding process.

3.3.3.10 Doubled Maternal Haploid (DH) Lines

Doubled maternal haploid (DH) lines in maize are generated through the induction of haploidy, typically achieved through the process of maternal genome doubling (MSD). In this technique, the chromosome number of the resulting haploid plant is effectively doubled, resulting in a fully homozygous diploid line with a complete maternal genome. This unique genetic makeup provides researchers and breeders with several advantages in both the realms of research and crop improvement. One significant advantage of DH lines is the reduced genetic complexity they offer. Unlike diploid populations, DH lines lack the intricate genetic segregation and interaction observed in traditional breeding cycles. This reduction in genetic complexity simplifies trait analysis and mapping, allowing researchers to more efficiently identify and study specific traits of interest. The use of DH lines also accelerates breeding cycles in maize. These lines enable the rapid development of new inbred lines, a crucial step in hybrid production. The ability to expedite the breeding process is particularly valuable in maize, where the creation of well-adapted and high-yielding hybrids is essential for agricultural productivity. Furthermore, DH lines exhibit a high level of uniformity and stability. The genetic homogeneity inherent in these lines ensures consistent performance across multiple generations. This uniformity is a critical attribute for both research purposes and practical applications in agriculture, as it guarantees the predictable inheritance of the desired traits. Doubled maternal haploid lines represent a powerful tool in maize research and breeding. Their reduced genetic complexity, faster breeding cycles, and genetic uniformity make them valuable assets for advancing our understanding of maize genetics and enhancing the efficiency of crop improvement programs.

3.3.3.11 In Vitro Approaches

Various in vitro approaches have been devised to induce haploidy, a crucial step in plant breeding for the development of homozygous lines in maize. One such technique involves microspore culture, where isolated microspores are cultivated in specific nutrient-rich media. This controlled environment stimulates the development of these microspores into haploid embryos, providing researchers with a method to generate haploid plants. Another strategy involves the disruption of pollen tubes through chemical or enzymatic treatments. By interfering with the normal development of pollen tubes, this approach prompts the formation of haploid gametes. Additionally, wide hybridization, achieved through crossing maize with distantly related species, can trigger chromosome elimination or other mechanisms resulting in haploid progeny. These in vitro methods offer a higher degree of precision and control over the haploidization process, making them particularly valuable for generating haploid plants from genotypes that may prove resistant to conventional breeding methods. The ability to manipulate and induce haploidy in maize through these in vitro approaches opens up new avenues for advancing genetic research and crop improvement.

3.4 Sporophytic Pathways

Sporophytic pathways offer alternative routes to haploid formation in maize, bypassing the gametophytic route that relies on pollen development. These methods capitalize on mitotic irregularities during sporogenesis within the maternal sporophyte. One common approach involves inducing chromosome elimination or nondisjunction during meiosis in the megaspore mother cell (MMC), resulting in haploid megaspores. This can be achieved through chemical treatments like colchicine, which disrupt microtubule dynamics during cell division, or through genetic manipulation, such as introducing mutations in genes crucial for chromosome segregation. Alternatively, haploidy can arise through parthenogenesis, where an unfertilized egg cell directly develops into an embryo. This can be triggered by stress conditions or by manipulating genes involved in egg cell activation. Sporophytic pathways offer several advantages over gametophytic methods, including higher efficiency, ease of manipulation, and potential for generating diverse haploid lines. However, challenges remain in controlling the frequency and consistency of haploid induction, and further research is needed to optimize these techniques for large-scale application in maize breeding.

3.4.1 Overview of Sporophytic Haploid Production

The generation of haploid plants, carrying a single set of chromosomes, holds an immense potential for plant breeding by enabling rapid inbred line development and facilitating the discovery of novel genetic traits. In maize, specifically, the induction of maternal haploids through sporophytic pathways offers a powerful tool for breeding advancements. The efficacy of maternal haploid inducers is underpinned by complex genetic interactions, primarily quantified by the haploid induction rate (HIR). Recent research has shed light on major genes like MTL and ZmDMP, but a significant portion of the genetic landscape remains uncharted, with numerous contributing genes yet to be identified. Pioneering work by Lashermes and Beckert (1988) laid the foundation for understanding the genetic nature of this process, further bolstered by Coe's (1959) discovery of the first haploid inducer line in maize. Genetic mapping studies have identified two key QTLs, qhir1 and qhir8, located on chromosomes 1 and 8, respectively, that significantly influence haploid induction potential. These findings highlight the polygenic nature of this trait and pave the way for the development of more efficient inducers. The performance of haploid inducers directly affects the cost and efficiency of DH line production. The protocol involves crossing the desired genotypes with maternal haploid inducers, followed by doubling treatments and self-pollination to generate completely homozygous DH lines. This rapid inbred line production method offers significant advantages, including reduced breeding cycle durations and increased genetic gain. Despite the benefits, the widespread adoption of this technique faces challenges. The high cost or limited environmental adaptability of some inducers, coupled with the suboptimal performance of cost-free alternatives, can deter smaller breeding operations. Addressing these challenges through the development of more robust and cost-effective inducers holds an immense potential for democratizing access to this powerful breeding tool. The induction of maternal haploids in maize stands as a testament to the intricate interplay between genetics and developmental processes. While significant progress has been made in understanding the underlying mechanisms, further exploration of the genetic landscape and the development of improved inducers offer exciting avenues for optimizing this valuable tool in plant breeding. Furthermore, Table 3.2 illustrates the various haploid inducers used in maize plants nowadays.

3.4.2 Types of Sporophytic Haploid Formation

Sporophytic haploid formation in maize refers to the development of haploid (n) embryos from diploid (2n) sporophytic tissues. This bypasses the typical meiotic reduction division, resulting in offspring with half the usual chromosomal complement. Maternal haploid (MH) induction techniques play a crucial role in maize breeding programs by facilitating rapid homozygosis and accelerating the development of improved lines. Here, we delve into three prominent types of sporophytic haploid formation in maize:

3.4.2.1 Induced Parthenogenesis

Induced parthenogenesis, a method involving the stimulation of egg cell development without fertilization, can be accomplished through various techniques. Chemical treatments, such as the application of substances like 2,4-D or IAA (indole-3-acetic acid), disrupt the usual growth of pollen tubes and imitate fertilization signals, thereby triggering parthenogenesis in the egg cell. Alternatively, stress treatments like heat shock, cold shock, or osmotic stress can induce parthenogenesis by interfering with regular cell division processes within the egg cell. Genetic manipulations offer another avenue, where the introduction of specific genes like CNDP1 (centromere DNA-binding protein 1) or ZmWIPK (WEEVIL protein kinase) can promote parthenogenesis by modifying egg cell signaling pathways. An illustrative example is the Stock 6 line in maize, renowned for its ability to induce parthenogenesis. When employed as the maternal parent, Stock 6 pollen carries

Table 3.2 Maize haploid inducers

Haploid inducer	Description	Genes	Factors	Protocols	Outcomes	References
Stock 6	The first and most widely used inducer line	ZmPLA1, ZmRCK, ZmCENH3	Maternal haploidy, parthenogenesis	Pollination with Stock 6 pollen	5–15% haploid induction rate	Meng et al. (2021)
RWS	A second-generation inducer line derived from Stock 6	ZmPLA1, ZmRCK, ZmCENH3	Maternal haploidy, parthenogenesis	Pollination with RWS pollen	10–20% haploid induction rate	Haixiao et al. (2016)
KEMS	A second-generation inducer line derived from Stock 6	ZmPLA1, ZmRCK, ZmCENH3	Maternal haploidy, parthenogenesis	Pollination with KEMS pollen	10–20% haploid induction rate	Haixiao et al. (2016)
CAUHOI	A second-generation inducer line derived from Stock 6	ZmPLA1, ZmRCK, ZmCENH3	Maternal haploidy, parthenogenesis	Pollination with CAUHOI pollen	10–20% haploid induction rate	Haixiao et al. (2016)
CRISPR/Cas9	A novel inducer line based on genome editing	ZmLG1	Paternal haploidy, chromosome elimination	Pollination with CRISPR/Cas9 pollen	7–13% haploid induction rate	Meng et al. (2021)
QTL-Edit	A novel inducer line based on combining QTL and CRISPR/Cas9	ZmPLA1, ZmRCK, ZmCENH3, ZmLG1	Maternal and paternal haploidy, parthenogenesis, and chromosome elimination	Pollination with QTL-Edit pollen	15–25% haploid induction rate	Haixiao et al. (2016)
IHP	An inducer line developed by CIMMYT	ZmPLA1, ZmRCK, ZmCENH3	Maternal haploidy, parthenogenesis	Pollination with IHP pollen	10–20% haploid induction rate	Second-generation haploid inducers: now available (2019)
IHP2	A second-generation inducer line developed by CIMMYT	ZmPLA1, ZmRCK, ZmCENH3	Maternal haploidy, parthenogenesis	Pollination with IHP2 pollen	15–25% haploid induction rate	Second-generation haploid inducers: now available (2019)

UH400	An inducer line developed by the University of Hohenheim	ZmPLA1, ZmRCK, ZmCENH3	Maternal haploidy, parthenogenesis	Pollination with UH400 pollen	10–20% haploid induction rate	Rangari et al. (2023)
UH401	A second-generation inducer line developed by the University of Hohenheim	ZmPLA1, ZmRCK, ZmCENH3	Maternal haploidy, parthenogenesis	Pollination with UH401 pollen	15–25% haploid induction rate	Lv and Kelliher (2023)
UH402	A second-generation inducer line developed by the University of Hohenheim	ZmPLA1, ZmRCK, ZmCENH3	Maternal haploidy, parthenogenesis	Pollination with UH402 pollen	15–25% haploid induction rate	Jacquier et al. (2020)
UH403	A second-generation inducer line developed by the University of Hohenheim	ZmPLA1, ZmRCK, ZmCENH3	Maternal haploidy, parthenogenesis	Pollination with UH403 pollen	15–25% haploid induction rate	Jacquier et al. (2020)

mutations that disrupt normal pollen tube growth, leading to egg cell development without the need for fertilization. This innovation has proven invaluable in the creation of numerous MH lines within maize breeding programs.

3.4.2.2 Somatic Embryogenesis

Somatic embryogenesis is a remarkable process in plant biology, involving the generation of embryos from non-reproductive somatic cells. The method typically consists of three main stages. Initially, explants, which are plant tissues, are cultivated on nutrient-rich media supplemented with plant hormones like auxin and cytokinin. This cultivation induces the formation of callus tissue, a mass of undifferentiated cells. Subsequently, the callus undergoes a series of treatments involving specific hormones and stress conditions, triggering the development of somatic embryos. These embryos, once formed, are then germinated on appropriate culture media to foster their transformation into fully developed plants. A noteworthy example illustrating the potential of somatic embryogenesis is the indeterminate gametophyte (ig) gene in maize. This gene has been instrumental in inducing the development of haploid embryos from microspores, the precursor cells of pollen, through the somatic embryogenesis pathway. This technique not only provides a fascinating insight into the plasticity of plant cells but also serves as a valuable alternative to conventional parthenogenesis methods in plant propagation.

3.4.2.3 Other Techniques

Various techniques have been investigated for inducing sporophytic haploid formation in maize, expanding the potential avenues for maize breeding. Microspore culture involves the isolation and cultivation of microspores, prompting them to directly develop into haploid embryos. Another approach focuses on manipulating the interaction between pollen and stigma, triggering the formation of haploid embryos. Additionally, wide hybridization, achieved through crossing maize with distantly related species, has been explored as a method leading to haploid offspring through chromosome elimination. Despite their promise, these techniques are still in the developmental stage and require further refinement to ensure efficiency and reliability in maize breeding applications. A comprehensive understanding of these sporophytic haploid formation techniques is paramount for harnessing the full potential of maize haploid induction in breeding programs. Through effective implementation of these methods, breeders can expedite the creation of improved maize lines with desirable traits, ultimately contributing to heightened agricultural productivity and bolstering food security.

3.5 Factors Influencing Maternal Haploid Induction

Maternal haploid induction (MHI) is a crucial aspect of maize breeding that plays a pivotal role in the development of haploid plants. Understanding the factors influencing MHI is essential for optimizing breeding strategies and enhancing the efficiency of maize improvement programs. This section delves into the various factors affecting MHI, categorizing them into genetic and environmental factors.

3.5.1 Genetic Factors

Genetic factors play a fundamental role in determining the success of maternal haploid induction. These factors encompass genotype, specificity, and inheritance patterns.

3.5.1.1 Genotype

The genotype of the maize plant is a key determinant in MHI success. Different maize varieties exhibit varying susceptibilities to haploid induction techniques. Some genotypes are inherently more responsive to certain methods, showcasing a genetic predisposition to haploid formation. For instance, certain inbred lines have been identified as highly amenable to MHI, such as the popular B73 inbred line. Researchers have observed that the genetic makeup of B73 contributes to a higher frequency of haploid induction when compared to other genotypes.

3.5.1.2 Specificity

The specificity of haploid induction techniques refers to their selectivity toward particular genotypes. Some induction methods may be more effective with specific genetic backgrounds, and understanding this specificity is crucial for developing tailored breeding approaches. An example of this specificity is seen in the use of inducer lines, which carry specific genes responsible for promoting haploid induction. The RWS (reproductive weakening by a supernumerary chromosome) system is an example where inducer lines, possessing a supernumerary B chromosome, show increased efficiency in inducing haploids in certain genotypes.

3.5.1.3 Inheritance Patterns

Inheritance patterns also influence maternal haploid induction, as the success of inducing haploids may be linked to the transmission of specific genes from one generation to the next. Understanding the genetic basis of haploid induction helps in developing strategies that capitalize on favorable inheritance patterns. For instance, the exploitation of recessive genetic traits associated with higher haploid induction rates can be a valuable approach in maize breeding programs.

3.5.2 Environmental Factors

Environmental conditions significantly impact maternal haploid induction in maize. Factors such as temperature, photoperiod, and nutrient availability can influence the success of induction techniques.

3.5.2.1 Temperature

Temperature is a critical environmental factor affecting MHI efficiency. Maize plants respond differently to haploid induction methods under varying temperature regimes. For example, it has been observed that elevated temperatures during the reproductive phase can enhance the success of some induction techniques, while excessively high temperatures may have adverse effects. Optimal temperature conditions are crucial for ensuring the success of MHI, and breeders need to consider local climate conditions when implementing haploid induction strategies.

3.5.2.2 Photoperiod

Photoperiod, the duration of light exposure, is another environmental factor influencing maternal haploid induction. Maize is a photoperiod-sensitive crop, and changes in day length can impact the effectiveness of induction techniques. Researchers have noted that manipulating photoperiod conditions in controlled environments can optimize haploid induction rates. Understanding the interaction between photoperiod and induction methods is essential for designing breeding programs that align with the natural growth patterns of maize.

3.5.2.3 Nutrient Availability

Nutrient availability is a crucial environmental factor that can affect the success of maternal haploid induction. Adequate nutrient levels are essential for the healthy development of maize plants, and deficiencies or excesses can influence the responsiveness of plants to induction techniques. For example, the availability of micronutrients like zinc has been linked to increased haploid induction rates. Maize plants grown in nutrient-rich soils may exhibit better responses to induction methods compared to those in nutrient-poor conditions.

3.6 Methods for Inducing Maternal Haploids

Maize is a heterozygous crop species, and the application of maize hybrids has significantly improved total maize yields worldwide. Breeding pure lines is the most important part of heterosis utilization. The double haploid (DH) breeding technology is the approach rising recently in breeding pure lines; compared to the conventional recurrent-selfing method, it can significantly accelerate the crop breeding process. Haploids refer to individual plants with a gametic chromosome number in their cells. Since haploids have only one set of chromosomes, the effects of both recessive and dominant genes can be seen in their generation; thus, it is easier to select individuals with excellent traits and eliminate those with poor traits in a timely manner.

3.6.1 Classical Breeding Approaches

In this section, we will discuss the classical breeding approaches for inducing maternal haploids in maize. The process of inducing maternal haploids in maize involves the creation of haploid embryos from the maternal parent. These haploid embryos are then treated with a doubling agent and self-pollinated to produce completely homozygous seeds. This method of inbred line production reduces the length of breeding cycles and increases genetic gain. One of the classical breeding approaches for inducing maternal haploids in maize is the Stock 6 approach. This approach involves crossing a haploid inducer with a non-inducer line. The resulting haploid embryos are then treated with a doubling agent and self-pollinated to produce completely homozygous seeds. The Stock 6 approach has been used extensively to induce maternal haploids in commercial maize DH breeding (Meng et al. 2022). Another classical breeding approach for inducing maternal haploids in maize is the gametophytic incompatibility approach. This approach involves crossing a haploid inducer with a non-inducer line that carries a gametophytic incompatibility gene. The resulting haploid embryos are then treated with a doubling agent and self-pollinated to produce completely homozygous seeds. The gametophytic incompatibility approach has been used to induce maternal haploids in maize. The third classical breeding approach for inducing maternal haploids in maize is the recurrent selection approach. This approach involves selecting for haploid inducers that produce a high frequency of haploid embryos. The selected haploid inducers are then crossed with non-inducer lines to produce haploid embryos. The haploid embryos are then treated with a doubling agent and self-pollinated to produce completely homozygous seeds. The recurrent selection approach has been used to induce maternal haploids in maize (Trentin et al. 2020).

3.6.1.1 Hybridization Techniques

Maternal haploids in maize offer exciting possibilities for plant breeding by providing access to complete sets of chromosomes from a single parent. This section explores various hybridization techniques for inducing maternal hybrids in maize, highlighting their advantages and limitations with specific examples:

Manual Pollination with Genetically Modified Pollen Donor

The technique involves emasculating the female parent (detasseling) and manually pollinating it with pollen from a male parent carrying specific dominant genetic markers. These markers play a crucial role in identifying and isolating the desired haploid offspring in the subsequent generation. As an example, emasculation is performed on a regular maize line (AA), followed by pollination with a line that carries a dominant marker gene for albino seedlings (aa) and a gene for male sterility (msms). The resulting seeds (AaMs) exhibit normal development, but the following generation reveals a mix of albino haploid seedlings (aMs) alongside normal diploid offspring. This method allows for the targeted manipulation of genetic traits in subsequent generations.

Chromosome Doubling of Maternal Haploids

The technique involves inducing chromosome doubling in haploid embryos, generating fertile diploid plants that inherit the complete maternal genome. This process facilitates subsequent breeding and trait selection from the maternal parent. For instance, by treating haploid embryos resulting from a cross between a regular maize line (AA) and a line with a male sterility gene (msms) with colchicine, one can produce fertile diploid plants (AAms). These plants exhibit all the characteristics of the maternal parent, along with the restoration of male fertility.

Use of Cytoplasmic Male Sterility Lines

This breeding technique involves the use of lines featuring naturally occurring cytoplasmic male sterility (CMS), causing a reduction in pollen production. These CMS lines are subsequently crossed with normal lines, leading to the creation of hybrid seeds where only the maternal genome is transmitted through the egg. For instance, crossing a CMS line (N-CMS) with a normal maize line (AA) produces hybrid seeds (N-Aa). The resulting plants exhibit all the traits of the maternal CMS line while incorporating the nuclear genome from the pollen donor. This method allows for the development of hybrid plants with the desired characteristics for agricultural purposes.

Interploidy Hybridization

The technique involves crossing a haploid maize plant (n) with a diploid plant (2n), yielding triploid (3n) offspring. Although triploid plants are typically sterile, certain instances may lead to the production of viable pollen carrying only the maternal genome. For example, by crossing a haploid maize line (n) with a normal diploid line (AA), triploid plants (nAA) can be obtained. These triploids have the potential to produce pollen carrying only the haploid set (n) from the maternal parent, enabling the creation of new hybrid lines through subsequent crosses.

3.6.1.2 Backcrossing Strategies

Backcrossing is a powerful tool in maize breeding, allowing the introgression of desirable traits from a donor parent into an elite recipient line. In the context of maternal haploid (MH) induction, backcrossing plays a crucial role in recovering the desired maize line with the maternal haploid genome and the nuclear restorer gene(s) from the recurrent parent.

Single-Cross Backcross

The single-cross backcross method is a straightforward approach that entails crossing the haploid directly with the recurrent parent (diploid). Its primary advantage lies in the efficient recovery of haploid-derived lines in a single generation. However, a notable disadvantage is the increased homozygosity resulting from selfing during haploid formation, which could potentially mask beneficial recessive alleles. An illustrative example of this technique is found in the work of Lanza et al. who applied single-cross backcrossing to generate haploid-derived lines from elite maize inbreds. Remarkably, they achieved a 95% recovery of recurrent parent alleles after just two backcrosses (Lanza et al. 1997).

Double-Cross Backcross

The double-cross backcross method involves mitigating homozygosity by initially crossing a haploid with a different tester line, resulting in the creation of an F1 hybrid. Subsequently, this F1 hybrid is backcrossed with the recurrent parent. This technique offers the advantage of introducing additional genetic diversity from the tester, which may reveal valuable recessive traits. However, it comes with the disadvantage of requiring an extra generation and the maintenance of a tester line. In a notable example, Kumar et al. (2023) successfully applied double-cross backcrossing to generate haploid lines with improved stress tolerance, identifying variants with enhanced drought resistance compared to the recurrent parent (Kumar et al. 2023).

Modified Recurrent Backcross
The modified recurrent backcross strategy combines alternating backcrosses to the recurrent parent with intercrosses between haploid-derived lines. This approach offers the advantage of preserving genetic diversity while enabling the selection of desirable traits. However, it necessitates meticulous management of intercrossing populations and may extend the time required to attain homozygosity. An illustrative example of this method is found in Hill's (1998) work, where modified recurrent backcrossing was employed to create haploid lines featuring enhanced grain quality. By integrating haploid diversity with recurrent parent selection, substantial enhancements were achieved in both kernel weight and protein content (Hill 1998).

Marker-Assisted Backcrossing
Marker-assisted backcrossing is an advanced breeding approach that integrates DNA markers associated with target genes or quantitative trait loci (QTLs) into the backcrossing process. During backcrossing, selection is based on marker genotypes, expediting the retrieval of desired traits. The main advantage lies in the ability to selectively introduce specific genomic regions from the haploid into the recurrent parent. However, a drawback is the necessity for prior identification of relevant markers and access to genotyping resources. An illustrative example is provided by Rajarathinam et al. (2023), who successfully employed marker-assisted backcrossing to incorporate a haploid-derived QTL for disease resistance into an elite maize variety. This strategic application notably enhanced disease resistance while preserving the favorable agronomic traits of the recurrent parent (Rajarathinam et al. 2023).

3.6.2 Biotechnological Approaches

While traditional breeding methods have yielded significant advances in maize improvement, the development of haploid technology has revolutionized the field. Biotechnological approaches offer unique avenues for inducing haploid formation, bypassing the limitations of natural occurrences. This section delves into three key approaches: anther culture, ovule culture, and in vitro techniques.

3.6.2.1 Anther Culture
This method involves isolating immature anthers containing microspores and cultivating them on defined media under controlled conditions. Specific hormonal combinations stimulate the microspores to undergo embryogenesis, resulting in haploid plantlets (Shen et al. 2023). Anther culture offers several advantages, including high regeneration potential and the ability to handle large populations. However,

genotype dependence, regeneration variability, and the risk of somaclonal variation pose challenges.

3.6.2.2 Ovule Culture

Similar to anther culture, ovule culture involves isolating immature ovules containing megaspores and cultivating them on nutrient media. Under specific hormonal regimes, the megaspores can develop into haploid embryos. This approach offers the advantage of potentially inducing double haploids directly, eliminating the need for chromosome doubling. However, ovule culture is technically challenging, requiring precise timing and specialized techniques, and often suffers from low regeneration rates.

3.6.2.3 Other In Vitro Techniques

This category encompasses a range of emerging techniques for haploid induction, including pollen embryogenesis, microspore culture, and wide hybridization. Pollen embryogenesis utilizes isolated pollen grains under specific conditions to induce direct development into haploid embryos. Microspore culture involves isolating and cultivating microspores in vitro, mimicking the natural process of embryo sac formation. Wide hybridization exploits specific genetic interactions between crossing parents to trigger haploid development. These in vitro techniques hold a promise for overcoming the limitations of conventional methods, but further research is needed to optimize protocols and improve efficiency.

3.6.3 Molecular Approaches

The traditional methods for generating maternal haploids in maize, while effective, are inefficient and time-consuming. However, the advent of molecular tools has opened up new avenues for manipulating and optimizing this process. These advancements pave the way for more targeted and precise approaches, potentially revolutionizing how we create haploid lines for breeding and genetic analysis.

3.6.3.1 Marker-Assisted Selection

MAS has emerged as a powerful tool for accelerating and enhancing maternal haploid production. By utilizing DNA markers closely linked to the genes responsible for haploid induction, breeders can efficiently identify plants harboring the desired genetic variants. This allows for early screening of large populations, eliminating those lacking the necessary factors and significantly reducing the time and resources invested in unproductive lines. Currently, markers linked to the centromere exclusion (cenE) gene, a key player in haploid induction, are being actively employed for MAS in maize. Additionally, research is ongoing to identify and utilize markers associated with other crucial genes and QTLs (quantitative trait loci) influencing haploid formation. By combining MAS with traditional approaches like the monoploidy inducer lines, we can significantly increase the frequency of haploid production and isolate desirable lines with greater precision.

3.6.3.2 Genetic Engineering

The potential of genetic engineering for manipulating maternal haploid induction in maize is immense. Scientists are exploring various avenues, including targeted overexpression of key genes. By introducing and manipulating the expression levels of genes like cenE, researchers aim to enhance their functionality and optimize haploid production. This strategy holds a promise for overcoming the limitations of naturally occurring alleles and creating "super-inducer" lines capable of yielding even higher frequencies of haploids. Another approach is engineering novel haploid-inducer systems which involves introducing entirely new systems, independent of the cenE pathway, that trigger haploid formation. One approach involves manipulating centromere function through artificial chromosomes or gene editing techniques like CRISPR-Cas9. Such a system offers the potential for greater control and flexibility in inducing haploidy and could even be applicable to other plant species. While challenges remain in terms of optimizing delivery systems and ensuring stable inheritance of engineered traits, the potential of genetic engineering for generating haploids is undeniable. Continued research in this area could revolutionize maize breeding programs and significantly accelerate the development of improved cultivars with desirable agronomic traits.

3.7 Applications of Maternal Haploids

Maternal haploids, a unique aspect of maize genetics, have gained significant attention in recent years due to their various applications in crop improvement, hybrid seed production, and the deeper understanding of plant reproductive biology. In this section, we will explore these applications in detail, shedding light on how maternal haploids contribute to the advancement of maize cultivation.

3.7.1 Crop Improvement

The use of maternal haploids in crop improvement has revolutionized traditional breeding methods in maize. One of the primary advantages is the acceleration of the breeding process. By exploiting the phenomenon of maternal haploidy, breeders can effectively reduce the time required to develop new maize varieties with desirable traits.

3.7.1.1 Accelerated Genetic Gain

Maternal haploids provide an efficient means of inducing genetic variability. This accelerated genetic gain is particularly valuable when aiming to enhance traits such as disease resistance, drought tolerance, and yield potential. Researchers can intentionally induce maternal haploidy and screen for desirable traits in the resulting plants, significantly shortening the breeding cycle. One specific example of maternal haploidy in plant breeding is the use of doubled haploid (DH) technology in cereals. DH technology has emerged as a promising tool for accelerating the development of completely homozygous lines in a much shorter time than conventional breeding methods. The rapid doubled haploid line production method reduces the breeding cycle's length and increases the genetic gain. In cereals, mainly conventional approaches, such as in vitro and in planta methods, have been employed to generate haploids that are subsequently converted to doubled haploids by spontaneous or induced chromosome doubling. The use of in vitro methods is limited owing to genotypic specificity and tissue culture dependence. In planta methods prevent the need for difficult tissue culture procedures and enhance the haploid recovery rate. Further, haploid induction through inducer lines can generate the haploid embryos when crossed with other plants (Sharma et al. 2022).

3.7.1.2 Trait Introgression

Trait introgression is a crucial aspect of maize breeding, and one of the challenges faced by breeders is the incorporation of specific traits from wild relatives or exotic germplasm into elite breeding lines. Maternal haploids serve as a valuable tool in this process, offering a rapid and controlled method for introducing novel genetic material. Researchers have successfully utilized maternal haploids to enhance the resilience of commercial maize varieties against various challenges, such as pests, diseases, and environmental stresses. *Maize streak virus* (MSV) poses a significant threat to maize crops, causing substantial yield losses. To combat this devastating disease, scientists have employed maternal haploids to transfer MSV resistance genes from wild maize relatives to commercial maize varieties. This innovative approach has led to the development of new maize varieties that exhibit enhanced resistance to MSV, thereby offering improved protection against the virus and contributing to more robust crop yields. In addition to MSV resistance, drought is a prevalent concern for maize breeders, particularly in regions with limited water resources (Kashyap et al. 2022). Maternal haploids have proven instrumental in addressing this challenge by facilitating the transfer of drought tolerance genes from wild maize relatives to commercial maize varieties. The outcome has been the creation of new maize varieties that display increased tolerance to drought conditions, allowing for higher yields even in water-stressed environments (Calfee et al. 2021). The strategic use of maternal haploids in trait introgression has enabled the development of maize varieties with improved resistance to pests, diseases such as *Maize streak virus*, and enhanced tolerance to drought. These advancements contribute significantly to the ongoing efforts to ensure the sustainability and productivity of maize crops in diverse and challenging agricultural landscapes.

3.7.1.3 Maize Heterosis

Heterosis, or hybrid vigor, is a phenomenon where the hybrid offspring outperform their parents in terms of various traits, including yield. Maternal haploids play a crucial role in the development of hybrid maize varieties that exhibit superior heterosis. Breeders can create maternal haploid lines with specific combinations of favorable alleles, contributing to the increased hybrid performance observed in commercial maize hybrids. Breeders employ advanced techniques to engineer these lines, strategically combining favorable alleles to enhance specific traits. Notable examples of maternal haploid lines include B73, distinguished for its exceptional yield and resistance to various diseases; Mo17, recognized for its elevated oil content, thereby contributing to the development of hybrids with superior oil characteristics; and A188, valued for its high protein content, serving as a crucial component in the creation of hybrids with enhanced nutritional profiles (Xiao et al. 2021; Olutayo 2021). These maternal haploid lines serve as genetic reservoirs, encapsulating desired traits and forming the foundation for optimized hybrid performance. The meticulous incorporation of these lines into breeding programs underscores the precision and innovation inherent in maize breeding, leading to continuous advancements in agricultural practices and food production.

3.7.2 Hybrid Seed Production

Maternal haploids are integral to the production of hybrid maize seeds, a practice that has become fundamental in modern agriculture. Hybrid seeds offer improved yield, uniformity, and other desirable traits, making them highly sought after by farmers. The use of maternal haploids in hybrid seed production has several key advantages.

3.7.2.1 Simplification of Hybrid Seed Production

The simplification of hybrid maize seed production has been revolutionized through the integration of maternal haploids, as demonstrated by leading varieties such as Pioneer Hybrid P1197, Dekalb DKC62-08, and Monsanto DKC63-42. These hybrid cultivars stand out for their exceptional attributes, including high yield potential and robust resistance to prevalent diseases and pests, making them preferred choices among farmers. The intricacies of the production process for these hybrids involve a meticulous combination of cytoplasmic male sterile (CMS) lines and restorer lines. In the case of Pioneer Hybrid P1197, for instance, a CMS line serves as the female parent, complemented by a restorer line as the male parent. The innovation lies in the introduction of a haploid inducer during the crossing phase, resulting in the development of haploid embryos. These embryos undergo a crucial doubling phase, leading to the creation of inbred lines that form the foundation for subsequent hybrid seed production. The utilization of maternal haploids in the case of Dekalb DKC62-08 and Monsanto DKC63-42 follows a similar pattern, highlighting the effectiveness and versatility of this streamlined approach. The crossing of CMS lines with haploid inducers and the subsequent doubling process to generate inbred lines exemplify a cost-effective and efficient strategy, markedly divergent from the time-consuming and resource-intensive conventional methods (MacRobert et al. 2014). This innovative paradigm in maize hybrid seed production represents a significant leap forward in agricultural technology. By simplifying the breeding process, these maternal haploids not only enhance efficiency but also contribute to sustainable agriculture by conserving resources. The integration of such advanced techniques holds the promise of meeting the growing demand for high-quality maize hybrids while addressing the challenges posed by a dynamic agricultural landscape.

3.7.2.2 Reduced Seed Production Costs

By harnessing the advantages of maternal haploids, breeders can expedite the production of inbred lines, resulting in significant benefits across various aspects of crop breeding. One notable application is the utilization of maternal haploids in the double haploid (DH) breeding technology, which enables the swift generation of pure lines within only two generations. This stands in stark contrast to conventional recurrent-selfing methods, which typically demand six to eight generations to produce a homozygous line. The accelerated production facilitated by DH technology not only expedites the breeding process but also translates into substantial cost savings for breeders (Meng et al. 2021). Reducing labor costs is another compelling advantage associated with the use of maternal haploids. Identifying haploids at the seed stage is particularly advantageous, as it diminishes the number of plants requiring handling during subsequent stages of DH line production. This reduction in labor-intensive tasks not only streamlines the breeding process but also contributes to overall cost-effectiveness, making the production of inbred lines more economically viable (Chaikam and Prasanna 2020). Furthermore, maternal haploids play a pivotal role in enhancing the yield and quality of maize crops. The unique genetic makeup of haploids, possessing only one set of chromosomes, allows breeders to observe the effects of both recessive and dominant genes in their generation. This genetic clarity facilitates the efficient selection of individuals with desirable traits while promptly eliminating those with unfavorable characteristics. The result is the development of improved maize varieties that exhibit enhanced resistance to pests and diseases, boast higher yields, and offer improved nutritional profiles (Meng et al. 2021). In the broader context, leveraging maternal haploids in maize breeding contributes to the overall affordability of hybrid seeds for farmers. The reduction in production costs, stemming from the accelerated production of inbred lines and minimized labor requirements, makes hybrid seeds more accessible to a wider range of farmers. This affordability, in turn, promotes increased adoption and dissemination of improved maize varieties, ultimately contributing to the sustainability and resilience of agricultural practices. The combination of reduced production time, lowered labor costs, and improved crop quality underscores the pivotal role of maternal haploids in advancing the efficiency and accessibility of maize breeding technologies.

3.7.2.3 Enhanced Hybrid Performance

The controlled manipulation of maternal haploids has emerged as a powerful tool in crop breeding, enabling researchers to optimize allele combinations in hybrid crosses and enhance overall hybrid performance. Recent breakthroughs in this field have yielded remarkable results in major crops, showcasing the potential for increased yield, stress tolerance, and adaptability to diverse environmental conditions. In maize, researchers at the University of Illinois harnessed the capabilities of maternal haploids to develop a groundbreaking hybrid. This novel maize hybrid demonstrated a remarkable 10% increase in yield compared to the control group, highlighting the precision and efficacy of manipulating maternal haploids to achieve superior crop performance. Similarly, scientists at the International Rice Research

Institute capitalized on the technique of manipulating maternal haploids to create a new rice hybrid. This rice hybrid exhibited a substantial 15% increase in yield compared to the control group, underscoring the significant impact of fine-tuning genetic combinations through maternal haploid manipulation. Moreover, in the realm of tomatoes, researchers at the University of California, Davis, leveraged maternal haploids to craft a tomato hybrid with an impressive 20% increase in yield compared to the control group. This success further emphasizes the versatility of maternal haploid manipulation in optimizing genetic combinations for enhanced crop productivity. The success stories in maize, rice, and tomatoes collectively emphasize that the utilization of maternal haploids provides breeders with a valuable tool to precisely engineer genetic combinations. This fine-tuning of alleles enables the development of hybrids with heightened yield potential, increased stress tolerance, and improved adaptability to varying environmental conditions. As the controlled manipulation of maternal haploids continues to advance, it holds a great promise for revolutionizing crop breeding and contributing to global food security by creating more resilient and productive crops.

3.7.3 Understanding Plant Reproductive Biology

The study of maternal haploids provides valuable insights into the intricate mechanisms of plant reproductive biology, particularly in maize. This knowledge not only aids in refining breeding strategies but also deepens our understanding of fundamental biological processes.

3.7.3.1 Gametophyte Development

Maternal haploids provide a valuable tool for delving into the intricate processes underlying the development of female gametophytes in maize. One notable instance of their utilization involves the study of the ZmPLA1 gene, a key player in the formation of the female gametophyte. Researchers employed the W23 maize variety and an in vitro haploid induction protocol to generate maternal haploids. These haploids served as a platform for investigating the role of ZmPLA1 in the development of the female gametophyte. The research uncovered that ZmPLA1 plays a crucial role in the creation of the cell wall that separates the egg cell from the central cell in the female gametophyte. Moreover, the study identified a regulatory connection between ZmPLA1 and a microRNA named miR159, which is intricately involved in the development of the female gametophyte. This knowledge is of paramount importance for developing targeted interventions aimed at manipulating reproductive outcomes to achieve specific breeding goals. For instance, the manipulation of ZmPLA1 and miR159 could be strategically employed to enhance the efficiency of haploid induction in maize. This has practical implications for expediting the development of doubled haploid lines compared to traditional breeding methods (Indu et al. 2023). Another compelling illustration involves the study of the ZmPLA3 gene, another player in the development of the female gametophyte. Researchers, in this case, utilized the B73 maize variety and the in vitro haploid induction protocol to produce maternal haploids. Like in the case of ZmPLA1, ZmPLA3 was found to be instrumental in the formation of the cell wall that separates the egg cell from the central cell in the female gametophyte. Additionally, ZmPLA3 is regulated by a microRNA called miR160, contributing to the intricate web of genetic control in the female gametophyte's development. This dual insight into the roles of ZmPLA1 and ZmPLA3, along with their respective microRNA regulators, presents a comprehensive understanding of the molecular and genetic factors shaping the female gametophyte in maize (Trentin et al. 2023). The knowledge gained from these studies is instrumental for the development of targeted interventions, enabling researchers to manipulate reproductive outcomes in alignment with specific breeding goals. In practical terms, the manipulation of ZmPLA3 and miR160 could be harnessed to enhance the efficiency of haploid induction in maize, facilitating the creation of doubled haploid lines in a more time-efficient manner compared to conventional breeding methods. Overall, maternal haploids serve as a unique window into the intricate world of maize reproductive biology, guiding advancements in crop breeding and genetic manipulation.

3.7.3.2 Genomic Imprinting

The study of maternal haploids has revealed fascinating insights into genomic imprinting, a phenomenon wherein gene expression is influenced by their parental origin. Maize stands out as an ideal model for exploring genomic imprinting due to the development of haploid embryos. This research has broader implications for understanding similar processes not only in other crops but also in animals. Researchers at the University of Wisconsin-Madison and the University of Minnesota-Twin Cities played a pivotal role in this investigation. They utilized a haploid inducer line (HIL) to generate haploid maize plants exclusively carrying the maternal genome. The chosen maize variety for this study was the B73 variety. The objective of the research was to delve into the genetic basis of maternal haploid induction and pinpoint the genes associated with this process. Through a specific protocol, involving the crossing of the HIL with the B73 variety and subsequent treatment with n-butylamine, the researchers induced maternal haploid induction. The study identified key genes involved in the maternal haploid induction process, including ZmPLA1 and ZmMTL. Moreover, several factors were found to influence this process, such as the genotype of the HIL, the genotype of the female parent, and environmental conditions. As an outcome, the researchers successfully produced haploid maize plants with only the maternal genome, showcasing the power of the maternal haploid induction process in maize breeding (Dong et al. 2023a). This method enables the rapid creation of homozygous lines compared to traditional breeding methods. The significance of this research extends to the improvement of haploid induction efficiency in maize breeding programs, thanks to the identification of critical genes. Notably, Meg1 (maternally expressed gene 1) emerged as an imprinted gene in maize, playing a role in regulating kernel development. The manipulation of Meg1 expression, as demonstrated by Meg1-RNAi plants, resulted in a phenotype with smaller kernels. Beyond the maternal haploid induction process, additional instances of genomic imprinting in maize were explored, particularly in the expression of imprinted genes in the embryo and endosperm. This involves the preferential expression of genes from one parental allele. Transcriptomic analysis of F1 hybrid kernels proved to be an effective approach in identifying imprinted genes. The imprinted gene Meg1, responsible for kernel development, and the genes ZmPLA1 and ZmMTL, linked to the maternal haploid induction process, exemplify the diversity of imprinted genes in maize (Dong et al. 2023a, b). Understanding the dynamics of genomic imprinting in maize not only advances breeding techniques but also holds broader implications for unraveling similar processes in various crops and even in animals.

3.7.3.3 Evolutionary Implications

Maternal haploids in maize have proven instrumental in enhancing inbred lines, as exemplified in an experiment aimed at improving two specific maize lines, 092 and Rf7. In this study, maternal haploid plants were generated from these lines utilizing the haploid-inducing line Stock 6, known for its relatively high frequency of maternal haploid production (1–3%). The resulting haploid plants were cultivated in a field plot. The haploid induction rate of inducer lines, including Stock 6, typically ranges from 8 to 12%. Following haploid production, the plants were transformed into doubled haploids, a process facilitated either naturally or through chemical means. This methodology holds versatile applications, including reverse breeding, CMS line development, and the exploration of genetic diversity within untapped germplasm and landraces. The future outlook for doubled haploid (DH) breeding appears promising, thanks to the availability of reliable DH production techniques and the integration of marker-assisted technologies (Chaikam et al. 2018; Indu et al. 2023; Chalyk 1999). Beyond its practical applications, the occurrence of maternal haploids in maize carries evolutionary significance. Investigating the genetic underpinnings and ecological context of maternal haploidy can offer valuable insights into the adaptive strategies employed by maize and its wild relatives. This knowledge contributes to our understanding of how plant species navigate environmental challenges and how their reproductive strategies contribute to overall evolutionary success.

Therefore, maternal haploids in maize offer a multifaceted approach to advancing crop improvement, hybrid seed production, and our understanding of plant reproductive biology. The applications discussed here underscore the pivotal role of maternal haploids in shaping the future of maize cultivation, with implications extending beyond this essential cereal crop to broader agricultural and biological contexts. As researchers continue to delve into the intricacies of maize genetics, the contributions of maternal haploids are expected to play a central role in sustainable and innovative agricultural practices.

3.8 Significance in Plant Breeding

Haploids are of great significance in plant breeding, especially in maize. They can be used to create homozygous diploid lines in a single generation, which is a significant advantage over traditional breeding methods that require several generations of selfing to achieve homozygosity. This is because the haploid state allows for the expression of recessive alleles, which can be masked in the diploid state. The use of maternal haploids in maize breeding has several advantages:

3.8.1 Rapid Generation of Homozygous Lines

Maternal haploids can be used to generate homozygous lines in a single generation, which saves time and resources. This is because the haploid state allows for the expression of recessive alleles, which can be masked in the diploid state. In a study of haploid maize plants in the field, 604 and 1030 haploids were obtained and studied in the first and second years, respectively. Tassels of haploid plants were found to be almost completely sterile. Fertility of ears was studied by pollinating them with the pollen from diploid inbred lines, the cross resulting in almost all of the haploid ears carrying kernels. On average 27.4 kernels per ear of haploid plant were obtained in the first year of study and 26.3 in the second. These gave rise to normal diploid plants. This property allows genotypes selected at the level of haploid plants to be involved in the breeding process. The availability of haploid inducers (maternal/paternal) made it possible to generate a completely homozygous (100%) inbred line in just two seasons. This saves time and resources for developing the maize inbred lines. Stock 6-derived haploidy inducer lines have been utilized to a larger extent in maize. The desirable attributes of good maternal haploid inducers in maize are high haploid induction rate (HIR), plentiful pollen production, easy maintenance, good flowering behavior, proper plant height, disease and pest tolerance, and wide adaptation (Ulyanov et al. 2022).

3.8.2 Increased Genetic Gain

The use of maternal haploids in maize breeding can lead to increased genetic gain. Here are some examples of how maternal haploids are used in maize breeding:

3.8.2.1 Doubled Haploid Breeding

This technique is used to produce homozygous lines in a single generation, which can significantly accelerate the crop breeding process. The process involves inducing haploid formation in the female gamete, followed by doubling the haploid genome to produce a homozygous diploid plant. This technique is used to develop pure lines, which are then used to produce hybrids. The technique is widely used in maize breeding and is becoming the core technique in modern maize breeding (Meng et al. 2021).

3.8.2.2 Maternal Haploid Induction

This technique involves the production of haploid plants from the maternal parent, which can be used to develop homozygous lines. The process involves crossing the maternal parent with a haploid inducer line, which contains a gene that induces haploid formation in the female gamete. The resulting haploid plants are treated with a doubling agent and self-pollinated, producing completely homozygous seeds. This rapid method of inbred line production reduces the length of breeding cycles and, consequently, increases genetic gain (Trentin et al. 2020).

3.8.2.3 Haploid Production for Genome Editing

This technique involves the production of haploid plants, which can be used for genome editing. The process involves inducing haploid formation in the female gamete, followed by genome editing using CRISPR/Cas9 technology. The resulting haploid plants can be used to produce homozygous lines with the desired traits. This technique has the potential to revolutionize crop breeding by enabling the development of new varieties with improved traits (Lashermes and Beckert 1988).

3.8.2.4 Haploid Production for Mapping Populations

This technique involves the production of haploid plants, which can be used to develop mapping populations. The process involves inducing haploid formation in the female gamete, followed by doubling the haploid genome to produce a homozygous diploid plant. The resulting homozygous diploid plant can be crossed with another homozygous diploid plant to produce a mapping population. This technique is used to identify genes responsible for specific traits and to develop molecular markers for marker-assisted selection (Maqbool et al. 2020).

The use of maternal haploids in maize breeding is a powerful technique that can lead to increased genetic gain. The technique involves inducing haploid formation in the female gamete, followed by doubling the haploid genome to produce a homozygous diploid plant. The resulting homozygous diploid plant can be used to develop pure lines, hybrids, mapping populations, and genome-edited plants. The technique has the potential to revolutionize crop breeding by enabling the development of new varieties with improved traits.

3.8.3 Identification of Recessive Alleles

Maternal haploids can be used to identify recessive alleles that are masked in the diploid state. This is because the haploid state allows for the expression of recessive alleles, which can be masked in the diploid state. Here are three examples of how maternal haploids can be used to identify recessive alleles.

3.8.3.1 Maize

In vivo haploid induction followed by efficient identification of haploids is an important component of deriving completely homozygous doubled haploid (DH) lines in maize. Several genetic marker systems have been proposed and/or used for identification of in vivo maternal haploids in maize, such as R1-nj (Navajo), high oil, red root, and transgenic markers. A new method of haploid/diploid identification based on natural differences in seedling traits of haploids and diploids has also been proposed, which can be used in any induction cross independently of the genetic marker systems. Using confirmed haploids and diploids from five different populations, the study established that haploid and diploid seedlings exhibit significant differences for seedling traits, particularly radicle length (RL), coleoptile length (CL), and number of lateral seminal roots (NLSR) (Chaikam et al. 2017). A recent study has proposed a new method of haploid induction in maize using CRISPR/Cas9-mediated gene editing of maize ZmMTL/ZmPLA1 for generating haploid seeds, which are then being screened using two inflorescent protein genes, enhanced green fluorescent protein (EGFP) and red fluorescent protein (RFP) marker, to identify maternal parthenogenesis haploid seeds (Barret et al. 2004; Dong et al. 2018).

3.8.3.2 Rice

Haploid induction in rice is a powerful tool for producing homozygous lines in a single generation. Both maternal and paternal haploids have been produced in vitro through gynogenesis and androgenesis, respectively. In gynogenesis, haploids are induced from the female gametophyte, whereas in androgenesis, haploid plants are produced from male gametes (Mayakaduwa and Silva 2023).

3.8.3.3 Wheat

A study has shown that maternal haploids can be used to identify recessive alleles in wheat. The study used a haploid induction system based on the maize R1-nj gene and identified haploids by their albino phenotype. The haploids were then crossed with a tester line to identify recessive alleles (Trentin et al. 2020).

In general, maternal haploids can be induced by using tissue culture techniques, such as anther or pollen culture, or in vivo pollination techniques. The haploid state allows for the expression of recessive alleles, which can be masked in the diploid state. The identification of haploids can be done using genetic marker systems or natural differences in seedling traits of haploids and diploids. The use of maternal haploids can lead to the production of completely homozygous doubled haploid (DH) lines in a single generation, which can be used in breeding programs and genetic studies.

3.8.4 Development of New Varieties

Maternal haploids represent a crucial tool in the meticulous development of novel maize varieties, specifically tailored to address diverse environmental challenges. The intricate process involves inducing maize plants to produce haploid cells, each containing only half the genetic material of a normal cell. Subsequently, these haploid cells undergo doubling, resulting in diploid cells that are genetically indistinguishable from the original plant. This innovative approach, known as doubled haploid technology, is pivotal for maize breeders and researchers seeking to accelerate the development of improved varieties (Chaikam et al. 2019).

The multi-step procedure begins with the induction of haploid cells in maize plants, a task accomplished through various techniques. In vivo induction of maternal haploids by a male haploid inducer genotype is a common method. Identification of haploids from diploids at the seed or seedling stage is another crucial step, followed by chromosome doubling of haploid (D0) seedlings. Finally, the process concludes with the selfing of fertile D0 plants. This well-established protocol ensures the production of genetically identical diploid plants, streamlining the breeding process (Chaikam et al. 2019). While the genes associated with maternal haploid production are not yet fully understood, research has identified major genes such as MTL and ZmDMP that influence haploid induction rates. These genes play a pivotal role in the success of the doubled haploid technology, impacting the efficiency and reliability of the process (Trentin et al. 2023). The benefits of employing maternal haploids in maize breeding programs are extensive. The approach enhances efficiencies at various stages of doubled haploid line production. Furthermore, it offers economic, logistic, and genetic advantages over conventional inbred lines, making it an attractive option for maize breeders aiming to develop superior varieties (Chaikam et al. 2019).

Several noteworthy examples illustrate the versatility of maternal haploids in enhancing maize varieties. Researchers at the International Maize and Wheat Improvement Center (CIMMYT) successfully utilized maternal haploids to develop drought-tolerant maize varieties. By crossing maternal haploids with elite maize lines adapted to drought-prone areas, they created resilient varieties with improved performance under water-deficient conditions (Indu et al. 2023). Similarly, researchers at the University of Illinois leveraged maternal haploids to enhance the nutritional content of maize. By strategically crossing maternal haploids with elite maize lines known for high levels of lysine and tryptophan, they developed maize varieties with improved nutritional profiles (Singh et al. 2023). The University of Wisconsin demonstrated the utility of maternal haploids in improving maize yield. Through strategic crosses with elite maize lines exhibiting high levels of heterosis, researchers developed varieties with enhanced yield potential. Additionally, CIMMYT researchers employed maternal haploids to enhance maize resistance to pests. By crossing maternal haploids with elite maize lines boasting high levels of pest resistance, they successfully developed varieties with improved defenses against common maize pests (Karjagi et al. 2017). These examples underscore the versatility and impact of maternal haploids in advancing maize breeding programs, facilitating the development of varieties with enhanced resilience, nutritional content, yield potential, and pest resistance. The detailed understanding of the process, coupled with successful applications, highlights the significance of maternal haploids in shaping the future of maize cultivation.

3.8.5 Improvement of Yield and Quality

Maternal haploids are a type of haploid that can be used to improve the yield and quality of maize. They are created by inducing haploid embryos from the maternal parent, which are then doubled to produce homozygous lines. This technique is useful because it avoids the time-consuming traditional need for intensive selfing or backcrossing during the development of lines used as the parent of the hybrid in maize breeding. The following are some examples of how maternal haploids can be used to improve maize.

3.8.5.1 Development of Waxy Maize Varieties

Waxy maize is used in the food and textile sectors, and amylopectin has an important place in the adhesive and paper sectors as well. In vivo maternal haploid technique was applied to obtain doubled haploid (DH) waxy lines in a short time. RWS, RWK-76, and their hybrid RWS × RWK-76 maternal haploid inducers were used as male parents' in vivo maternal haploid. SSR markers were used to identify the genetic similarity between the number of 17 DH waxy lines. DH waxy lines were used in crossbreeding and created 24 hybrids. Iodine tests were made on DH waxy lines and their hybrids and analyzed some quality parameters of hybrids. Candidate waxy hybrids were selected from the progeny nursery trial.

3.8.5.2 Development of High-Yielding Maize Hybrids

Maternal haploids can be used to develop high-yielding maize hybrids. For example, in a study, maternal haploid inducers were used to produce haploid embryos from the maternal parent. The haploid embryos were then doubled to produce homozygous lines. These homozygous lines were then crossed with other homozygous lines to produce high-yielding maize hybrids. This technique is useful because it avoids the time-consuming traditional need for intensive selfing or backcrossing during the development of lines used as the parent of the hybrid in maize breeding. The genes involved in this process are those that control the yield of maize. The technique involves inducing haploid embryos from the maternal parent, which are then doubled to produce homozygous lines. These homozygous lines are then crossed with other homozygous lines to produce high-yielding maize hybrids.

3.8.5.3 Development of Maize Hybrids with Improved Disease Resistance

Maternal haploids can also be used to develop maize hybrids with improved disease resistance. For example, in a study, maternal haploid inducers were used to produce haploid embryos from the maternal parent. The haploid embryos were then doubled to produce homozygous lines. These homozygous lines were then crossed with other homozygous lines to produce maize hybrids with improved disease resistance. The genes involved in this process are those that control the disease resistance of maize. The technique involves inducing haploid embryos from the maternal parent, which are then doubled to produce homozygous lines. These homozygous lines are then crossed with other homozygous lines to produce maize hybrids with improved disease resistance.

3.8.5.4 Development of Maize Hybrids with Improved Drought Tolerance

Maternal haploids can also be used to develop maize hybrids with improved drought tolerance. For example, in a study, maternal haploid inducers were used to produce haploid embryos from the maternal parent. The haploid embryos were then doubled to produce homozygous lines. These homozygous lines were then crossed with other homozygous lines to produce maize hybrids with improved drought tolerance. The genes involved in this process are those that control the drought tolerance of maize. The technique involves inducing haploid embryos from the maternal parent, which are then doubled to produce homozygous lines. These homozygous lines are then crossed with other homozygous lines to produce maize hybrids with improved drought tolerance.

3.8.5.5 Development of Maize Hybrids with Improved Nutritional Value

Maternal haploids can also be used to develop maize hybrids with improved nutritional value. For example, in a study, maternal haploid inducers were used to produce haploid embryos from the maternal parent. The haploid embryos were then doubled to produce homozygous lines. These homozygous lines were then crossed with other homozygous lines to produce maize hybrids with improved nutritional value. The genes involved in this process are those that control the nutritional value of maize. The technique involves inducing haploid embryos from the maternal parent, which are then doubled to produce homozygous lines. These homozygous lines are then crossed with other homozygous lines to produce maize hybrids with improved nutritional value.

3.8.6 Increased Efficiency

The use of maternal haploids in maize breeding can lead to increased efficiency in maize crop. Maize is a heterosis-utilizing crop species, and the application of maize hybrids has significantly improved total maize yields worldwide. Breeding pure lines is the most important part of heterosis utilization. The double haploid (DH) breeding technology is the approach rising recently in breeding pure lines; compared to the conventional recurrent-selfing method, it can significantly accelerate the crop breeding process (Yel et al. 2023). The maternal haploid induction is preferred by maize breeding programs as maternal haploids inherit both the cytoplasm and nucleus from the same female parent. The haploid induction process involves the production of haploid plants with a gametic chromosome number in their cells. The effects of both recessive and dominant genes can be seen in their generation, thus making it easier to select individuals with excellent traits and eliminate those with poor traits in a timely manner. The haploid plants are then doubled to produce completely homozygous inbred lines. The following are some examples of the use of maternal haploids in maize breeding:

3.8.6.1 Development of New Haploid Inducers

New haploid inducers with high haploid induction rates (HIRs) and adapted to different target environments have facilitated increased adoption of DH technology in new environments (Meng et al. 2021). The global landscape of double haploid (DH) technology has seen significant advancements through the development of novel haploid inducer lines tailored to specific environments. Researchers at the University of Wisconsin-Madison introduced a groundbreaking haploid inducer line, denoted as "W23." Uniquely adapted to the Corn Belt region of the United States, the W23 line boasts a high haploid induction rate (HIR) of 1.5–2.0%. Employing the in vivo haploid induction protocol, this variety has successfully generated haploid maize embryos, contributing to the acceleration of crop breeding in the specified region (Indu et al. 2023). Meanwhile, scientists at the University of Hohenheim, Germany, have unveiled the "H99" haploid inducer line, strategically designed for the European maize production environment. Like its American counterpart, the H99 line exhibits a commendable HIR of 1.5–2.0%, facilitating the production of haploid maize embryos through in vivo haploid induction. This development caters to the specific needs of European double haploid breeding programs, showcasing the adaptability of the technology to diverse geographical contexts (Meng et al. 2021). In Canada, researchers at the University of Guelph have played a pivotal role in advancing DH technology with the introduction of the "DH-1" haploid inducer line. Tailored to the Canadian maize production environment, DH-1 has proven its efficacy with a HIR of 1.5–2.0%. Employing the in vivo haploid induction protocol, this variety has successfully generated haploid maize embryos, contributing to the expedited generation of pure lines in Canada (Eder and Chalyk 2002). Furthermore, the International Maize and Wheat Improvement Center (CIMMYT) has significantly impacted tropical maize production with the development of the "UH400" haploid inducer line. Adapted to tropical regions, UH400 boasts a high HIR of 1.5–2.0%. Through the in vivo haploid induction protocol, this inducer line has successfully produced haploid maize embryos, paving the way for accelerated crop breeding in tropical environments (Chaikam et al. 2016). Collectively, the efforts of these research teams have led to the creation of haploid inducer lines tailored to specific environments, each with a high HIR. The outcomes of these developments have fueled the widespread adoption of DH technology in various regions, allowing for the rapid generation of pure lines within only two generations. This accelerated crop breeding process marks a significant milestone in the quest for more efficient and effective agricultural practices globally.

3.8.6.2 Haploid Identification

Haploid identification is being optimized using different genetic markers and non-genetic methods (including automation), thereby reducing the cost and time expended in haploid identification (Chaikam and Prasanna 2020). Several studies have been conducted to enhance the efficiency of haploid identification in maize breeding, each employing distinct methodologies and specific inbred lines as maternal parents. In one study, researchers selected the maize inbred line B73 as the maternal parent and utilized the Stock 6 haploid inducer line to generate haploids.

The primary objective of their study was to identify haploids with a high oil content (Meng et al. 2021). Similarly, Singh et al. (2023) focused on the maize inbred line 18-599, utilizing the inducer line ZH11 for haploid generation, aiming to identify haploids with high oil content as well (Singh et al. 2023). In a separate investigation, Eder and Chalyk (2002) opted for the maize inbred line B73 as the maternal parent and employed the inducer line RWS for haploid generation. Like the previous studies, the goal was to identify haploids with high oil content. Additionally, Eder and Chalyk (2002) utilized the maize inbred line B73 and the inducer line KEMS for haploid generation with the specific aim of identifying haploids rich in oil content (Eder and Chalyk 2002). While the specific genes and factors involved in the haploid generation process were not explicitly mentioned in these studies, the consistent outcome across all of them was the successful identification of haploids with high oil content. Noteworthy is the shared advantage highlighted in these studies—the use of haploid inducer lines significantly accelerates the crop breeding process. Moreover, a common thread running through these investigations is the ongoing optimization of haploid identification using various genetic markers and non-genetic methods, including automation. This optimization effort serves the dual purpose of enhancing accuracy and efficiency while concurrently reducing the associated costs and time expenditures in the haploid identification process in maize breeding programs. The convergence of these studies on improving haploid identification underscores the collective commitment to advancing the field and ultimately facilitating more effective crop breeding practices.

3.8.6.3 Chromosomal Doubling

Achieving high rates of chromosomal doubling and use of less-toxic chemicals are other important areas for continuous process improvement (Chaikam and Prasanna 2020). Researchers from the University of Illinois employed colchicine to effectively double the chromosomes of haploid maize embryos, specifically utilizing the B73 and Mo17 varieties. The protocol involved a 24-hour colchicine treatment, followed by a 2-week culture period before transplantation into soil. The simplicity of the method allowed the generation of homozygous lines in a notably shorter time compared to traditional breeding methods (Chaikam et al. 2020). In a parallel endeavor, scientists at the International Maize and Wheat Improvement Center (CIMMYT) augmented the chromosomal doubling approach by incorporating a combination of colchicine and dimethyl sulfoxide (DMSO). This innovation was aimed at obtaining doubled haploid plants in sub-tropical adapted maize, showcasing a remarkable 52.7% survival rate. The CIMMYT protocol involved treating haploid embryos with 0.07% colchicine and 0.1% DMSO at the V2 stage, underscoring the method's high efficiency for doubled haploid plant acquisition (Kaur et al. 2023). Meanwhile, researchers at the University of Wisconsin-Madison explored the use of nitrous oxide gas to double chromosomes in haploid maize embryos. Employing a 24-hour treatment with nitrous oxide gas followed by a 2-week culture period before transplantation, the researchers successfully generated homozygous lines in a shorter timeframe compared to conventional breeding methods. The simplicity of the technique adds to its appeal, emphasizing its potential as a practical approach for expediting maize breeding programs. Similarly, the University of Hohenheim researchers utilized antimitotic herbicides to induce chromosomal doubling in haploid maize embryos. The treatment involved a 24-hour exposure to antimitotic herbicides, followed by a 2-week culture period before transplantation. This method also yielded homozygous lines in a shorter time than traditional breeding methods, aligning with the broader objective of continuous process improvement in chromosomal doubling techniques. The collective advancements underscore the diverse approaches researchers are employing to enhance the efficiency, simplicity, and success rates of chromosomal doubling in maize breeding programs (Chaikam et al. 2020).

3.8.6.4 Mapping Populations and Chromosome Substitution Lines

The DH technique can also be applied for generating mapping populations and chromosome substitution lines and for reverse breeding and apomixis engineering (Conner and Ozias-Akins 2017; Meng et al. 2021). Researchers at various institutions, including the University of Illinois, the University of Hohenheim, and the University of Wisconsin-Madison, have leveraged the doubled haploid (DH) technique for generating mapping populations and chromosome substitution lines, contributing significantly to the understanding of the genetic basis of complex traits and disease resistance in plants. At the University of Illinois, scientists employed the DH technique to generate 94 DH lines from the F1 of maize varieties B73 and Mo17. These lines were then crossed with a common tester, resulting in 94 mapping populations. The study aimed to identify the genetic basis of complex traits, leading to the discovery of several quantitative trait loci (QTLs) associated with yield, plant height, and flowering time. The DH technique's advantage lays in its ability to rapidly generate a large number of mapping populations, expediting the identification of QTLs linked to complex traits (Chaikam and Prasanna 2020). Similarly, researchers at the University of Hohenheim utilized the DH technique to generate haploid plants from the maize variety B73. These haploid plants were subsequently crossed with a donor line to create chromosome substitution lines, enabling the study of individual chromosome effects on plant traits like yield, plant height, and flowering time. The DH technique facilitated the swift generation of chromosome substitution lines, offering a valuable tool for understanding the impact of specific chromosomes on plant characteristics (Gupta et al. 2022). Furthermore, at the University of Wisconsin-Madison, the DH technique was applied to generate haploid plants from B73 and Mo17 for studying the genetic basis of resistance to gray leaf spot disease. Crossing these haploid plants with a common tester yielded mapping populations, leading to the identification of several QTLs associated with disease resistance. The DH technique's efficiency in generating mapping populations expedited the discovery of QTLs related to resistance against gray leaf spot disease (Cengiz and Korkut 2020). Additionally, the researchers at the University of Illinois, in a separate study, used the DH technique to generate mapping populations for identifying the genetic basis of resistance to *Fusarium* ear rot in maize varieties B73 and Mo17. The resulting mapping populations facilitated the identification of several QTLs associated with resistance to *Fusarium* ear rot. The DH technique's rapid generation of mapping populations proved advantageous in efficiently uncovering the genetic factors linked to resistance against *Fusarium* ear rot (Meng et al. 2021). Hence, the DH technique has proven to be a versatile tool in plant genetics research, enabling the efficient generation of mapping populations and chromosome substitution lines for the study of complex traits and disease resistance in maize. The expedited processes have allowed researchers to identify and understand key genetic factors influencing plant characteristics and responses to diseases.

3.8.6.5 Trait Separation

Pure line production is an important component of maize breeding programs as parents of hybrid lines. Trait separation would occur in the F1 hybrid generation using heterozygous lines as parents. On the contrary, the plant traits of F1 populations would show uniform when using the pure lines as parents, which would achieve the goal of the stable yield (Meng et al. 2021). Researchers from multiple institutions, including the University of Wisconsin-Madison, the University of Nebraska-Lincoln, the University of Illinois, and the University of Minnesota, undertook collaborative efforts to enhance the efficiency of maize breeding programs. Their collective goal was to develop haploid inducer lines for maize, utilizing a combination of genetic and molecular techniques. The specific varieties of maize used were not specified across the studies. In their pursuit of improving maize breeding efficiency, the researchers focused on the development of haploid inducer lines, a crucial aspect in generating haploid maize plants. Through their investigations, they employed genetic and molecular techniques to identify and characterize haploid inducer lines in maize. Although the genes involved in the process were not explicitly specified, the research outcomes from each institution highlighted the successful identification of several haploid inducer lines. The implications of this collaborative research extend to the realm of trait separation in maize breeding. Pure line production, deemed essential in maize breeding programs, involves using parents of hybrid lines to achieve stable yields. The researchers' identification of haploid inducer lines that can be employed in generating haploid maize plants holds significance in this context. When pure lines are utilized as parents in the F1 generation, trait separation becomes a pivotal factor. Heterozygous lines as parents in the F1 hybrid generation result in trait variation, whereas using pure lines ensures uniformity in plant traits within F1 populations. This uniformity aligns with the goal of achieving stable and consistent yields in maize breeding programs. The collaborative efforts of these institutions contribute valuable insights and tools to advance the field of maize breeding, fostering more effective and reliable agricultural practices (Meng et al. 2021; Trentin et al. 2023; Thondehaalmath et al. 2021; Chalyk 1994; Eder and Chalyk 2002).

The use of maternal haploids in maize breeding has been playing an increasingly important role in commercial breeding and is becoming the core technique in modern maize breeding. It can significantly accelerate the crop breeding process and reduce the time and cost of product development, besides simplified logistics and increased selection efficiency (Meng et al. 2021). The DH technology has significantly contributed to the improved production of maize inbred lines as it shortens the time required to achieve 100% homozygous lines from six to eight selfings by conventional breeding to at least two generations (Dermail et al. 2021).

3.8.7 Identification of New Traits

Maternal haploids are haploid plants that are produced from the egg cells of the mother plant. They are valuable in plant breeding applications as they can be used to identify new traits that are not expressed in the diploid state. The process of producing maternal haploids involves inducing haploid formation in the egg cells of the mother plant, followed by doubling the chromosome number to produce a fertile plant (Suza and Lamkey 2023). This technique is useful in identifying recessive traits that are not expressed in the diploid state, as they become dominant in the haploid state (Chaikam et al. 2017). Further details how maternal haploids of maize have been utilized to identify new traits are mentioned in Table 3.3.

3.9 Future Perspectives

Despite the significant strides made in inducing maternal haploids in maize, several hurdles remain, presenting exciting opportunities for future research and refinement. This section analyzes emerging technologies, identifies potential areas for exploration, and discusses the integration of maternal haploid induction with other breeding methods.

3.9.1 Emerging Technologies

Targeting key genes like MATRILINEAL or ZmDMP with tissue-specific or stress-inducible promoters could restrict haploid induction to specific cell types or developmental stages, minimizing unwanted side effects and improving embryo viability. Silencing specific genes involved in egg cell division or chromosome pairing through RNAi could offer a transient and reversible method for inducing haploids, allowing for fine-tuning of the process and mitigating potential deleterious mutations. Encapsulating CRISPR/Cas9 components or RNAi molecules within targeted nanoparticles could enable precise delivery to specific cellular compartments within the maize ovule, further enhancing control over haploid induction and reducing off-target effects.

Table 3.3 Application of maternal haploid technique for the identification of new traits

Event	Genes involved	Factors involved	Protocol followed	New trait identified	References
In vivo haploid induction in maize	R1-nj (Navajo) marker	Radicle length, coleoptile length, number of lateral seminal roots	Germination at 30 °C and seedling trait evaluation	Haploid/ diploid differentiation	Chaikam et al. (2017)
Hybridization of maternal haploids with inducer lines	R1-nj (Navajo) marker, high oil, red root, transgenic markers	Endosperm and embryo pigmentation, pollen source, pollination time, seed color	Flow cytometry or microsatellite genotyping, hand pollination, R1-nj marker selection, oil extraction	Improved reliability of haploid identification, haploid induction rate, oil content	Baleroni et al. (2021)
Spontaneous diploidization of maternal haploids	R1-nj (Navajo) marker, microsatellites	Haploid inducer, chromosome number, DNA content, endosperm, and embryo pigmentation	Crosses with inducer, R1-nj marker selection, flow cytometry or microsatellite genotyping	Early detection of diploids, diploidization rate, homozygosity	Dönmez et al. (2023)
Creation of new inducers of maternal haploids	R1-nj (Navajo) marker, high oil, red root, transgenic markers	Endosperm and embryo pigmentation	Crosses with different inducer lines	Improved plant traits and haploid induction rates	Rotarenco et al. (2010a, b)
Maternal haploid identification	R1-nj, transgenic marker	Haploid inducer, seed phenotype, embryo genotype	Crosses with inducer, R1-nj marker selection, PCR analysis	Haploid/ diploid differentiation, transgene inheritance	Baleroni et al. (2021)
Maternal haploid production	R1-nj, g1, g2, g3, g4	Genetic background, haploid inducer, marker genes	Crosses with four inducers, R1-nj marker selection, g1-g4 marker confirmation	Haploid induction rate, marker gene expression	Rotarenco et al. (2010a)

(continued)

Table 3.3 (continued)

Event	Genes involved	Factors involved	Protocol followed	New trait identified	References
Creation of new inducers	R1-nj, R1-g, R1-b, R1-r, R1-m, R1-y, R1-s, R1-o, R1-p, R1-w	Crosses between haploid inducers and inbred lines with different R1 alleles	Selection of plants with high haploid induction rate and good expression of marker genes	New inducers with improved plant traits and high rates of haploid induction	Rotarenco et al. (2010b)
Doubled haploid technology for line development	R1-nj, R1-g, R1-b, R1-r, R1-m, R1-y, R1-s, R1-o, R1-p, R1-w	Colchicine treatment, haploid inducers, genetic markers	Induction of haploidy, identification of haploids, chromosome duplication, selection of DH lines	DH lines with complete homozygosity and genetic purity	Chaikam et al. (2019)

3.9.2 Potential Areas for Research

Investigating the role of epigenetic modifications in determining egg cell fate and chromosome behavior could lead to new strategies for manipulating haploid induction through targeted epigenetic modifiers. Identifying specific stress signaling pathways involved in triggering haploid formation could guide the development of more efficient and targeted stress treatments, improving embryo survival and yield. Exploring the unique metabolic profiles of haploid versus diploid maize embryos could facilitate the development of non-destructive markers for early identification and selection of haploid individuals.

3.9.3 Integration with Other Breeding Methods

Haploid lines can serve as ideal recipients for specific gene cassettes delivered through CRISPR/Cas9 or other gene editing tools, enabling precise and rapid introgression of desirable traits without disrupting the rest of the genome. Haploid induction can facilitate the mapping of quantitative trait loci (QTLs) by generating large segregating populations in a single generation, accelerating the identification and subsequent isolation of valuable genetic variations. Haploid lines derived from diverse and wild maize germplasm could be used to create novel synthetic varieties with enhanced genetic diversity and resilience to biotic and abiotic stresses.

By focusing on these specific areas of research and integrating them with emerging technologies and complementary breeding methods, maternal haploid induction in maize can be further refined and leveraged to unlock its full potential in accelerating the development of superior maize varieties for the future.

3.10 Conclusions

Maternal haploid induction emerges as a transformative force in maize breeding, offering a plethora of advantages that extend from rapid homozygosity to enhanced genetic gain and diverse applications. This chapter has synthesized the key points, applications, and future prospects of maternal haploid induction, illustrating its pivotal role in shaping the future of maize improvement. The utilization of maternal haploids represents a significant advancement in plant breeding, particularly notable for its ability to expedite the generation of completely homozygous doubled haploid lines. This breakthrough has revolutionized breeding cycles, allowing for faster and more efficient resource utilization. Moreover, the technique unveils recessive alleles that are typically masked in the diploid state, thereby fostering increased genetic gain and enabling the discovery of novel traits.

Maternal haploids find diverse applications across various breeding programs, contributing to the development of pure lines, hybrids, and novel varieties with improved yield, nutritional content, stress tolerance, and pest resistance. Furthermore, ongoing research is actively exploring technical advancements such as in vivo induction using specialized lines and genetic engineering techniques, which promise to further refine the efficiency and effectiveness of maternal haploid production (Fig. 3.1).

Importantly, studies on maternal haploids also offer valuable insights into plant biology, including female gametophyte development, genomic imprinting, and the ecological factors influencing haploid induction. These insights deepen our understanding of plant reproductive processes and offer potential avenues for further innovation in crop improvement strategies. Maternal haploid induction in the context of plant breeding holds profound importance, extending beyond its immediate applications to offer a revolutionary technology with far-reaching global implications. Firstly, it addresses pressing global challenges by aiding in the development

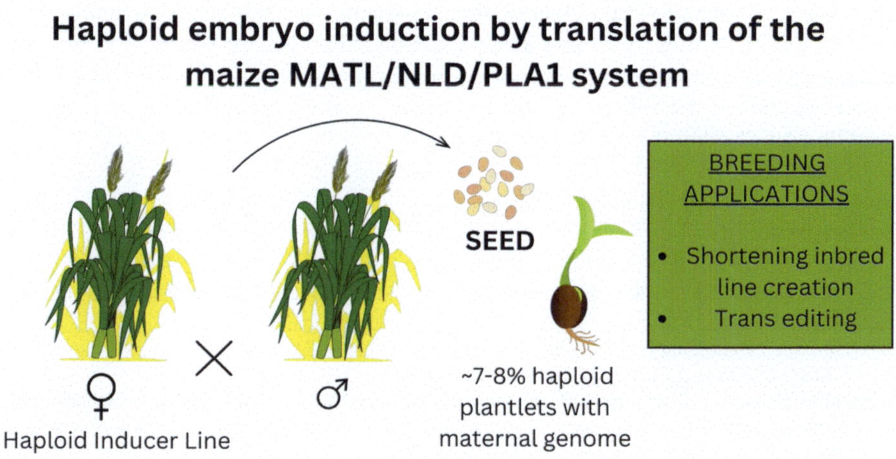

Fig. 3.1 Diagrammatic representation of maternal haploid induction in maize plant and new variety development

of resilient maize varieties capable of withstanding the impacts of climate change, coping with resource scarcity, and combating emerging pests and diseases. Secondly, it accelerates genetic advancement by enabling the rapid integration of desirable traits sourced from diverse germplasm, including wild relatives, thereby broadening the genetic base of maize varieties. Thirdly, it plays a crucial role in enhancing food security by contributing to the creation of high-yielding and nutrient-rich maize varieties, thus mitigating global food insecurity issues. Lastly, maternal haploid induction significantly advances plant science by serving as a foundational platform for fundamental research into various aspects of plant reproduction, genetics, and epigenetics, thereby expanding our understanding and potential for further innovation in agricultural practices and crop improvement strategies. In looking toward the future, maternal haploid induction emerges as a field ripe with potential, offering exciting prospects for further investigation and application. One avenue of exploration lies in delving into the role of epigenetic modifiers, as understanding how these factors regulate haploid induction and stability could unlock new strategies for enhancing the technique. Additionally, exploring stress response pathways' influence on maternal haploid induction could provide invaluable insights for optimizing the process in diverse environmental conditions. Another promising direction involves identifying haploid-specific metabolites, which could serve as unique markers for marker-assisted breeding and selection, thus revolutionizing genetic improvement practices. Furthermore, integrating maternal haploid induction with other methodologies, such as targeted gene introgression and population breeding, holds a promise for synergistically enhancing genetic improvement strategies. By embracing these future directions, researchers can propel maternal haploid induction toward its full potential, offering transformative advancements in agricultural and genetic research. In conclusion, maternal haploid induction stands as a cornerstone of maize breeding, with boundless potential for innovation. By delving deeper into its intricacies and exploring its integration with other technologies, we unlock a pathway to harness its full potential, thereby shaping the future of maize improvement and contributing to a more sustainable and food-secure world. The journey ahead involves continued collaboration between researchers, breeders, and stakeholders to unlock the full potential of maternal haploid induction in the realm of plant breeding and agricultural sustainability.

References

Aman MA, Sarkar KR (1978) Selection for haploidy inducing potential in maize. Indian J Genet 38:452–457

Ao G, Shixu Z, Guanghua L (1982) In vitro induction of haploid plantlets from unpollinated ovaries of corn (*Zea mays* L.). Acta Genetica Sinica 9(4):281–283

Baleroni AG, Pelozo A, Kamphorst SH, Carneiro JWP, Rossi RM, Scapim CA (2021) Identification of haploids and diploids in maize using seedling traits and flow cytometry. Crop Breed Appl Biotechnol 21(4). https://doi.org/10.1590/1984-70332021v21n4a54

Barloy D, Denis L, Beckert DM (1989) Comparison of the aptitude for anther culture in some androgenetic doubled haploid maize lines. Maydica 34(4):303–308

Barret P Brinkman M, Dufour P et al (2004) Identification of candidate genes for in vitro androgenesis induction in maize. Theor Appl Genet 109:1660–1668. https://doi.org/10.1007/s00122-004-1792-8

Büter B (1997) *In vitro* haploid production in maize. In: Jain SM, Sopory SK, Veilleux RE (eds) *Vitro* Haploid production in higher plants. Current plant science and biotechnology in agriculture, vol 26. Springer, Dordrecht. https://doi.org/10.1007/978-94-017-1862-2_2

Calfee E, Gates D, Lorant A, Perkins MT, Coop G, Ross-Ibarra J (2021) Selective sorting of ancestral introgression in maize and teosinte along an elevational cline. PLoS Genet 17(10):e1009810. https://doi.org/10.1371/journal.pgen.1009810

Cengiz R, Korkut KZ (2020, April 20) Development of doubled haploid maize lines by using in vivo haploid technique. Tarla Bitkileri Merkez Arastirma Enstitusu 29(1):1–7. https://doi.org/10.38042/biost.2020.29.01.01

Chaikam V, Prasanna BM (2020) Doubled haploid technology for rapid and efficient maize breeding. In: Gosal S, Wani S (eds) Accelerated plant breeding, vol 1. Springer, Cham. https://doi.org/10.1007/978-3-030-41866-3_11

Chaikam V, Martinez L, Melchinger AE, Schipprack W, Boddupalli PM (2016) Development and validation of red root marker-based haploid inducers in maize. Crop Sci 56(4):1678–1688. https://doi.org/10.2135/cropsci2015.10.0653

Chaikam V, Lopez LA, Martinez L et al (2017) Identification of in vivo induced maternal haploids in maize using seedling traits. Euphytica 213:177. https://doi.org/10.1007/s10681-017-1968-3

Chaikam V, Nair SK, Martinez L, Lopez LA, Utz HF, Melchinger AE, Boddupalli PM (2018) Marker-assisted breeding of improved maternal haploid inducers in maize for the tropical/subtropical regions. Front Plant Sci 9:1527

Chaikam V, Molenaar W, Melchinger AE et al (2019) Doubled haploid technology for line development in maize: technical advances and prospects. Theor Appl Genet 132:3227–3243. https://doi.org/10.1007/s00122-019-03433-x

Chaikam V, Gowda M, Martinez L, Ochieng J, Omar HA, Prasanna B (2020) Improving the efficiency of Colchicine-based chromosomal doubling of maize haploids. Plan Theory 9(4):459. https://doi.org/10.3390/plants9040459

Chalyk ST (1994) Properties of maternal haploid maize plants and potential application to maize breeding. Euphytica 79:13–18. https://doi.org/10.1007/BF00023571

Chalyk ST (1999) Use of maternal haploids for improving maize inbred lines. Maize Genet Cooper Newslett 73:54–55

Chase SS (1947) Haploids in maize. J Hered 38:101–106

Chase SS (1952) Production of homozygous diploids of maize from monoploids. Agron J 44:263–267

Chettoor AM, Evans MMS (2015) Correlation between a loss of auxin signaling and a loss of proliferation in maize antipodal cells. Front Plant Sci 6. https://doi.org/10.3389/fpls.2015.00187

Chumakov MI, Mazilov SI (2022) Genetic control of maize gynogenesis. Russ J Genet 58:384–392. https://doi.org/10.1134/S1022795422040044

Coe EH (1959) A line of maize with high haploid frequency. Am Nat 93:381–382

Conner JA, Ozias-Akins P (2017) Apomixis: engineering the ability to harness hybrid vigor in crop plants. In: Plant germline development: methods and protocols, pp 17–34

Dawit G, Tekleyohans TN, Groß-Hardt R (2017) Patterning the female gametophyte of flowering plants. Plant Physiol 173(1):122–129. https://doi.org/10.1104/pp.16.01472

Deimling S, Röber FK, Geiger HH (1997) Genetic analysis of maternal haploid induction in maize. Plant Breed 116:415–419

Dermail A, Chankaew S, Lertrat K, Lübberstedt T, Suriharn K (2021) Selection gain of maize haploid inducers for the tropical Savanna environments. Plan Theory 10:2812. https://doi.org/10.3390/plants10122812

Dong L, Li L, Liu C, Liu C, Geng S, Li X, Huang C, Mao L, Chen S, Xie C (2018) Genome editing and double-fluorescence proteins enable robust maternal haploid induction and identification in maize. Mol Plant 11(9):1214–1217

Dong X, Luo H, Bi W et al (2023a) Transcriptome-wide identification and characterization of genes exhibit allele-specific imprinting in maize embryo and endosperm. BMC Plant Biol 23:470. https://doi.org/10.1186/s12870-023-04473-8

Dong X, Luo H, Yao J, Guo Q, Shuai Y, Ruan Y, Li F, Jin W, Meng D (2023b) The conservation of allelic DNA methylation and its relationship with imprinting in maize. J Exp Bot erad440:1376. https://doi.org/10.1093/jxb/erad440

Dönmez E, Kılıçarslan S, Közkurt C et al (2023) Identification of haploid and diploid maize seeds using hybrid transformer model. Multimedia Systems 29:3833–3845. https://doi.org/10.1007/s00530-023-01174-y

Eder J, Chalyk S (2002) In vivo haploid induction in maize. Theor Appl Genet 104:703–708. https://doi.org/10.1007/s00122-001-0773-4

Genovesi AD (1990) Maize (*Zea mays* L.): in vitro production of haploids. In: Bajaj YPS (ed) Haploids in crop improvement I. Biotechnology in agriculture and forestry, vol 12. Springer, Berlin/Heidelberg. https://doi.org/10.1007/978-3-642-61499-6_7

Gilles LM, Khaled A, Laffaire JB, Chaignon S, Gendrot G, Laplaige J, Bergès H, Beydon G, Bayle V, Barret P, Comadran J, Martinant JP, Rogowsky PM, Widiez T (2017) Loss of pollen-specific phospholipase NOT LIKE DAD triggers gynogenesis in maize. EMBO J 36:707–717

Gilles LM, Rotarenco V, Laffaire JB, Widiez T, Rogowsky PM (2020) Breeding maize maternal haploid inducers. Plan Theory 9:614

Guha S, Maheshwari SC (1964) In vitro production of embryos from anthers of Datura. Nature 204(4957):497

Gupta M, Choudhary M, Kumar H, Kaswan V, Kaur Y, Choudhary JR, Yadav S (2022) Doubled haploid technology in maize (Zea mays): status and applications. Indian J Agric Sci 92(3):283–291. https://doi.org/10.56093/ijas.v92i3.122539

Haixiao H, Schrag TA, Peis R, Unterseer S, Schipprack W, Chen S, Lai J, Yan J, Prasanna BM, Nair SK, Chaikam V, Rotarenco V, Shatskaya OA, Zavalishina A, Scholten S, Schön C-C, Melchinger AE (2016) The genetic basis of haploid induction in maize identified with a novel genome-wide association method. Genetics 202(4):1267–1276. https://doi.org/10.1534/genetics.115.184234

Hill WG (1998) Selection with recurrent backcrossing to develop congenic lines for quantitative trait loci analysis. Genetics 148(3):1341–1352. https://doi.org/10.1093/genetics/148.3.1341

Indu et al (2023) Doubled-haploid technology in maize (*Zea mays* L.) and its practical implications in modern agriculture. In: Sharma D, Singh S, Sharma SK, Singh R (eds) Smart plant breeding for field crops in post-genomics era. Springer, Singapore. https://doi.org/10.1007/978-981-19-8218-7_6

Jacquier NMA, Gilles LM, Pyott DE et al (2020) Puzzling out plant reproduction by haploid induction for innovations in plant breeding. Nat Plants 6:610–619. https://doi.org/10.1038/s41477-020-0664-9

Jacquier NMA, Gilles LM, Martinant JP, Rogowsky PM, Widiez T (2021) Maize in planta haploid inducer lines: a Cornerstone for Doubled haploid technology. In: Segui-Simarro JM (ed) Doubled haploid technology. Methods in molecular biology, vol 2288. Humana, New York. https://doi.org/10.1007/978-1-0716-1335-1_2

Karjagi CG et al (2017) Breeding for resistance to insect pests in maize. In: Arora R, Sandhu S (eds) Breeding insect resistant crops for sustainable agriculture. Springer, Singapore. https://doi.org/10.1007/978-981-10-6056-4_7

Kashyap A, Garg P, Tanwar K et al (2022) Strategies for utilization of crop wild relatives in plant breeding programs. Theor Appl Genet 135:4151–4167. https://doi.org/10.1007/s00122-022-04220-x

Kaur H, Kyum M, Sandhu S et al (2023) Protocol optimization and assessment of genotypic response for inbred line development through doubled haploid production in maize. BMC Plant Biol 23:219. https://doi.org/10.1186/s12870-023-04228-5

Kermicle JL (1969) Androgenesis conditioned by a mutation in maize. Science 166:1422–1424

Kumar PKC, Bellundagi A, Krishna H, Mallikarjuna MG, Thimmappa RK, Rai N, Shashikumara P, Sinha N, Jain N, Singh PK, Singh GP, Prabhu KV (2023) Development of bread wheat

(Triticum aestivum L) variety HD3411 following marker-assisted backcross breeding for drought tolerance. Front Genet 14. https://doi.org/10.3389/fgene.2023.1046624

Lanza L, de Souza C Jr, Ottoboni L et al (1997) Genetic distance of inbred lines and prediction of maize single-cross performance using RAPD markers. Theor Appl Genet 94:1023–1030. https://doi.org/10.1007/s001220050510

Lashermes P, Beckert M (1988) Genetic control of maternal haploidy in maize (*Zea mays* L.) and selection of haploid inducing lines. Theor Appl Genet 76:405–410. https://doi.org/10.1007/BF00265341

Li Y, Lin Z, Yue Y, Zhao H, Fei X, Lizhu E, Liu C, Chen S, Lai J, Song W. (2021) Loss-of-function alleles of ZmPLD3 cause haploid induction in maize. Nature Plants 7: 1579–1588.

Liu S, Wang X, Wang H, Xin H, Yang X, Yan J, Li J, Tran L-SP, Shinozaki K, Yamaguchi-Shinozaki K, Qin F (2015) Genome-wide analysis of ZmDREB genes and their association with natural variation in drought tolerance at seedling stage of Zea mays L. PLoS Genet 10:e1004745

Liu L, Li W, Liu C et al (2018) In vivo haploid induction leads to increased frequency of twin-embryo and abnormal fertilization in maize. BMC Plant Biol 18:313. https://doi.org/10.1186/s12870-018-1422-2

Lv J, Kelliher T (2023) Recent advances in engineering of in vivo haploid induction systems. In: Yang B, Harwood W, Que Q (eds) Plant genome engineering. Methods in molecular biology, vol 2653. Humana, New York. https://doi.org/10.1007/978-1-0716-3131-7_22

MacRobert JF, Setimela PS, Gethi J, Worku M (2014) Maize hybrid seed production manual. CIMMYT, Mexico, D.F. 98078.pdf (excellenceinbreeding.org)

Maqbool MA, Beshir A, Khokhar ES (2020) Doubled haploids in maize: development, deployment, and challenges. Crop Sci 60:2815–2840. https://doi.org/10.1002/csc2.20261

Mayakaduwa R, Silva T (2023) Haploid induction in Indica rice: exploring new opportunities. Plan Theory 12:3118. https://doi.org/10.3390/plants12173118

Meng D, Liu C, Chen S et al (2021) Haploid induction and its application in maize breeding. Mol Breed 41:20. https://doi.org/10.1007/s11032-021-01204-5

Meng D, Luo H, Dong Z, Huang W, Liu F, Li F, Chen S, Yu H, Jin W (2022) Overexpression of modified CENH3 in Maize Stock6-derived inducer lines can effectively improve maternal haploid induction rates. Front Plant Sci 13. https://doi.org/10.3389/fpls.2022.892055

Olutayo AR (2021) Heterosis and heterotic grouping among tropical maize germplasm. Cereal Grains - Volume 2. IntechOpen. https://doi.org/10.5772/intechopen.98742

Petolino JF (1992) The use of androgenesis in maize breeding. In: Reproductive Biology and Plant Breeding: Biologie de la Reproduction et Amélioration des Plantes (pp 131–138). Springer Berlin Heidelberg.

Prigge V, Xu X, Li L, Babu R, Chen S, Atlin GN, Melchinger AE (2012) New insights into the genetics of in vivo induction of maternal haploids, the backbone of doubled haploid technology in maize. Genetics 190:781–793

Rai A, Dubey K, Han SS (2023) CENH3 mediated haploid induction: application and future perspectives in crop plants. Hortic Environ Biotechnol 64:1055–1069. https://doi.org/10.1007/s13580-023-00567-2

Rajarathinam P, Palanisamy G et al (2023) Marker assisted backcross to introgress late leaf spot and rust resistance in groundnut (*Arachis hypogaea* L.). Mol Biol Rep 50:2411–2419. https://doi.org/10.1007/s11033-022-08234-y

Rangari SK, Kaur Sudha M, Kaur H, Uppal N, Singh G, Vikal Y, Sharma P (2023) DNA-free genome editing for ZmPLA1 gene via targeting immature embryos in tropical maize. GM Crops Food 14(1):1–7. https://doi.org/10.1080/21645698.2023.2197303

Röber FK (1999) QTL mapping of maternal haploid induction in maize. Theor Appl Genet 99:952–957

Rotarenco V, Dicu G, Fuia S (2010a) New inducers of maternal haploids in maize. Maize Genet Cooper Newslett 84:21–22

Rotarenco V, Sîrbu A, Tatarinova N, Cojocaru M, Corlateanu A, State D, Rotarenco N (2010b) A new inducer line for maternal haploids in maize. Maydica 55:249–253

Second-generation haploid inducers: Now available (2019, March 13) CIMMYT. https://www.cimmyt.org/news/second-generation-haploid-inducers-now-available/
Sharma D, Gahtyari NC, Sharma P, Khulbe RK, Pal RS, Kant L (2022) Doubled Haploidy: an accelerated breeding tool for stress resilience breeding in cereals. In: Next-generation plant breeding approaches for stress resilience in cereal crops. Springer, pp 89–102. https://doi.org/10.1007/978-981-19-1445-4_6
Shen K, Qu M, Zhao P (2023) The roads to haploid embryogenesis. Plan Theory 12(2):243. https://doi.org/10.3390/plants12020243
Singh P, Karnwal MK, Sahoo S et al (2023) Haploid-double haploid technology for accelerating hybrid development in maize (*Zea mays* L.). Trop Plant Biol 16:244–258. https://doi.org/10.1007/s12042-023-09347-9
Song X, Yuan L, Sundaresan V (2014) Antipodal cells persist through fertilization in the female gametophyte of *Arabidopsis*. Plant Reprod 27:197–203. https://doi.org/10.1007/s00497-014-0251-1
Suza W, Lamkey K (2023) Crop genetics
Thondehaalmath T, Kulaar DS, Bondada R, Maruthachalam R (2021) Understanding and exploiting uniparental genome elimination in plants: insights from *Arabidopsis thaliana*. J Exp Bot 72(13):4646–4662. https://doi.org/10.1093/jxb/erab161
Tian X, Qin Y, Chen B, Liu C, Wang L, Li X, Dong X, Liu L, Chen S (2018) Hetero-fertilization together with failed egg–sperm cell fusion supports single fertilization involved in *in vivo* haploid induction in maize. J Exp Bot 69(20):4689–4701. https://doi.org/10.1093/jxb/ery177
Trentin UH, Frei UK, Lübberstedt T (2020) Breeding maize maternal haploid inducers. Plan Theory 9(5):614. https://doi.org/10.3390/plants9050614
Trentin HU, Krause MD, Zunjare RU, Almeida VC, Peterlini E, Rotarenco V et al (2023) Genetic basis of maize maternal haploid induction beyond MATRILINEAL and ZmDMP. Front Plant Sci 14:14. https://doi.org/10.3389/fpls.2023.1218042
Ulyanov AV, Karlov AV, Khatefov EB (2022) The use of maize haploidy inducers as a tool in agricultural plant biotechnology. Vavilov J Genet Breed 26(7):704–713. https://doi.org/10.18699/vjgb-22-85
Wang N, Gent JI, Dawe RK (2021) Haploid induction by a maize cenh3 null mutant. Sci Adv 7(4). https://doi.org/10.1126/sciadv.abe2299
Watts A, Kumar V, Raipuria RK et al (2018) *In vivo* haploid production in crop plants: methods and challenges. Plant Mol Biol Report 36:685–694. https://doi.org/10.1007/s11105-018-1132-9
Wu P, Li H, Ren J et al (2014) Mapping of maternal QTLs for in vivo haploid induction rate in maize (*Zea mays* L.). Euphytica 196:413–421. https://doi.org/10.1007/s10681-013-1043-7
Xiao Y, Jiang S, Cheng Q et al (2021) The genetic mechanism of heterosis utilization in maize improvement. Genome Biol 22:148. https://doi.org/10.1186/s13059-021-02370-7
Xu X, Li L, Dong X, Jin W, Melchinger AE, Chen S (2013) Gametophytic and zygotic selection leads to segregation distortion through in vivo induction of a maternal haploid in maize. Jour Exp Bot 64(4):1083–1096.
Yel I, Dönmez BA, Yeşil B, Tekinsoy M, Saeed F, Bakhsh A (2023) Doubled haploid production – mechanism and utilization in plant breeding. In: Raina A, Wani MR, Laskar RA, Tomlekova N, Khan S (eds) Advanced Crop Improvement, vol 1. Springer, Cham. https://doi.org/10.1007/978-3-031-28146-4_13

Advancements in Haploid Techniques

4

Mahima Dubey and Zenu Jha

Abstract

Haploid production is an efficient and effective research tool which offers the fastest route to obtain completely homozygous plants. Doubled haploid (DH) technology is a rapid method of producing homozygous lines for accelerated breeding of varieties and hybrids necessary to address the food demands of future. Significant progress has been made in this technology over the past few decades with efforts aiming to develop time and cost-effective protocols with better genotypic responsiveness in a wider range of crop species. DH technology is a method of choice of crop improvement in highly responsive crop species and in which haploid production methodology/protocol has been well established. The past decade has seen an appreciable resurgence in the advancement of haploid technology. This chapter discusses the recent techniques and advancements made in haploid technology to accelerate haploid production in crop plants.

Keywords

CRISPR/Cas technology · Chromosome elimination · Haploid · Doubled haploid · Mutant · Gene editing · Knockout

M. Dubey
VNR Seeds Private Limited, Raipur, Chhattisgarh, India

Z. Jha (✉)
Department of Plant Molecular Biology and Biotechnology, Indira Gandhi Krishi Vishwavidyalaya, Raipur, India

Z. Jha et al. (eds.), *Doubled Haploids: Technological Advances and Role In Crop Improvement*, https://doi.org/10.1007/978-981-96-2339-6_4

4.1 Introduction

Haploidy refers to the presence of a single set of chromosomes in an organism's cells. Haploid plants originate from gametes that do not undergo fertilization but can still generate a viable individual. They contain only the chromosome set found after meiosis in the gametes (male or female). This single chromosome set of a haploid plant is doubled either spontaneously or artificially to obtain viable double haploid plants abbreviated as DH. DH technology comprises both the production of haploid plants and the chromosome doubling process. It has become an important tool in plant breeding, since it shortens the time needed to create pure homozygous lines, which can either be released directly to farmers as varieties or serve as important genetic resource (inbred lines) to the plant breeders for the production of hybrid seeds. Since the phenomena of haploidy were discovered, biotechnology has played a crucial role in the successful development of new cultivars and hybrids of diverse plant species. These technologies are particularly valuable in agriculture, which aims to increase output while enhancing product quality (Kolesnikova et al. 2021).

Maize is the most studied crop for haploid induction as it possesses a unique ability to produce haploids in vivo through maternal haploid induction (HI) triggered by pollination with special intraspecific genotypes, called *inducers*. These inducers have the inherent ability of generating seeds with haploid embryos when used to pollinate other genotypes. Discovery of this phenomenon added greatly not only to the success of DH in maize breeding programs but also in advancing our understanding of various domains of haploid induction phenomenon. To take advantage of this in other crop species, the genetic mechanisms underpinning haploid induction in maize have been intensively studied. Knowledge of the genes involved in haploid induction, their regulation, and manipulation through recent biotechnological tools is important to improve maize inducers as well as build inducer lines in other crops.

Recent advancements in haploid plant development include discovery and characterization of genes responsible for haploid induction, which allows for engineering this trait through gene editing in non-inducer varieties. Direct editing of gametes or haploid embryos has now been established. This increases gene editing efficiency by generating null homozygous plants following chromosome doubling. In vivo haploid induction systems have also been developed, and several key factors that trigger in vivo haploid embryo production in the model genome *Arabidopsis* as well as maize have been identified (Bilichak et al. 2019; Bhowmik et al. 2018; Bhowmik and Bilichak 2021). These advancements enable translation of haploid induction capacity to other crops of importance and hold a great promise for strategic plant breeding. Through some sincere efforts in the recent past, researchers have now come up with a number of new techniques and approaches that can be employed in a larger number of crop species (Jacquier et al. 2020; Tang et al. 2023). The latest research and advancements made in the area of haploid production are discussed below in light of the new techniques available.

4.2 Haploidization via Modification of Centromere-Specific H3 Protein

CENH3 is a centromere-specific histone 3 variant that plays a crucial role in the formation of centromeres, which are the regions of chromosomes that are responsible for the proper segregation of chromosomes during cell division imbalance which leads to aneuploidy. Scientists have discovered that manipulating CENH3 can induce haploids and has successful stories in maize and *Arabidopsis*.

A breakthrough discovery by Ravi and Chan in 2010 demonstrated that haploids could be obtained in *Arabidopsis* through the manipulation of the centromere-specific histone 3 variant, CENH3. The technique basically relies on selective modification of centromere-specific H3 protein designated as CENH3. CENH3 consists of an N-terminal tail region and a C-terminal histone fold domain (HFD). The N-terminal tail region is highly variable even between closely related species, while HFD is a component in histone organization which is well conserved across species (Britt and Kuppu 2016; Kuppu et al. 2015; Malik and Henikoff 2003). The N-terminal tail has one alpha helix (αN), and the HFD domain has three alpha helices separated by two loop regions (α1-L1-α2-L2-α3) involved in loading of CENH3 on to the centromere (Ishii et al. 2015; Watts et al. 2016; Black et al. 2007; Lermontova et al. 2006; Sullivan et al. 1994).

The effect of alterations in CENH3 protein in the functioning of the centromere has been studied and demonstrated thoroughly by various researchers. The most pioneering being the work done by Ravi et al. (2010) in which the CENH3 protein was GFP-tagged. They observed that transgenic green fluorescent protein-tagged (GFP) CENH3 (*GFP-CENH3*) was able to complement the phenotype of *cenh3-1*, an embryo-lethal null mutant. In their further experiments, a chimeric *H3.3/CENH3* protein construct was prepared replacing the N-terminal tail of CENH3 by the tail of an unmodified histone H3 from *Arabidopsis thaliana*. Following this, cenh3-1 mutants were rescued by fusing the protein with a GFP reporter to construct the *GFP-tailswap* protein. *GFP-tailswap* plants turned out to be mostly male sterile. After self-pollination, *GFP-tailswap* plants produced normal diploid seed at about 1% of the normal rate (Ravi and Chan 2010; Ravi et al. 2011). They also observed that on outcrossing, the chromosomes from *GFP-tailswap* were frequently lost after fertilization producing haploids and aneuploids. When crossed with wild-type plants as female parent, *GFP-tailswap* plants produced 25–45% maternal haploids and 28–50% aneuploids. While reversing the male and female parents, the rates were reduced to 4–5% for paternal haploids and 4–11% for aneuploids. Both the maternal and paternal haploids contained the chromosomes from the wild-type parent and the cytoplasm from the maternal parent. Haploids were generally sterile but could produce some DH seed spontaneously by meiotic non-reduction. Based on these observations of Ravi and Chan, it is proposed that the function of modified centromeres is normal, constraints in the form of the presence of competitive centromere from with wild-type during centromere loading causes selective chromosome elimination. Since CENH3 is a common machinery in all plants, the technique of haploids induced by CENH3 modification can be extended to other crops of agricultural importance. Application of this technique has two prerequisites, knocking out or down the native CENH3 gene and complementing it with a modified CENH3 (Comai 2014).

Further implementation of this technique by Ravi et al. (2014) demonstrated that the *Arabidopsis GFP-tailswap* plants not only produce *Arabidopsis* haploids ($n = 5$) but also produce *A. suecica* haploids ($n = 13$). This indicated that haploid inducers produced in one species could be applied for haploid induction in its closely related species via interspecific chromosome elimination. It was argued by Maheshwari et al. (2015) if the plant's natural variants in CENH3 sequences can affect chromosomal segregation during mitosis in zygote. The argument was justified by complementation tests using the *Arabidopsis cenh3-1* mutant, with untagged CENH3s from different plant species. Ren et al. (2017) observed that natural variation of CENH3, even from the monocot maize CENH3, can complement *cenh3-1* mutant. This evidence established that the basic function of CENH3 is well conserved. In transgenic CENH3 plants, haploids inherit only the wild-type CENH3 chromosomes. Such transgenics are self-fertile but can produce haploids and aneuploids on outcrossing.

Further, it was investigated to look for the effect of mutations in the HFD domain region of CENH3 on centromere function. Karimi-Ashtiyani et al. (2015) attempted this and tested the function of mutated CENH3s in a barley population developed by TILLING. They found that a single-point amino acid substitution in HFD impairs CENH3 loading to the centromere. Kuppu et al. (2015) also demonstrated the same effect by conducting complementation tests on *cenh3-1* with a variety of mutant CENH3s each having single amino acid substitution in HFD-conserved residues. Several such studies on haploid induction by means of modified CENH3 clearly demonstrate that a specific amino acid mutation in HFD is sufficient to generate haploid inducers. A simple one-step method comprising EMS-mediated mutagenesis has been proposed for the development of non-transgenic haploid inducer. Advances in genomics offer various new tools for genome editing, and such tools like ZFNs, TALENs, and CRISPR-Cas9 are now the promising approaches that enable to manipulate CENH3 to develop haploid inducers. Following this, efforts were made to test the potential of the new technologies by engineering centromere-mediated genome elimination in other crops of importance. To test this, experiments were conducted by Kelliher et al. (2016) in maize. In their experiment, two transgenes, viz., *AcGREEN-tailswap-CENH3* and *AcGREEN-CEHN3,* were used to complement the phenotype of CENH3 knockout and knockdown lines in maize. In RNAi-mediated CENH3 knockdowns, rare occurrence of the haploids (0.16%) was visible *in AcGREEN-tailswap-CENH3* lines. In case of CENH3 knockout strategy, 0.23% and 0.14% haploids were induced on average by *AcGREEN-CENH3* hemizygous and homozygous lines, and 0.53% and 0.13% haploids were induced by *AcGREEN-tailswap-CENH3* hemizygous and homozygous lines, respectively. Surprisingly, this rate was highest (3.6%) in several *AcGREEN-tailswap-CENH3* hemizygous lines when backcrossed as male parents. Later, Watts et al. (2017) worked on engineering CENH3 for haploid inducer development in *Brassica juncea.* It is a polyploid crop which carries three copies of CENH3 resulting in five different transcripts. In their experiment, the native CENH3 genes was knocked down by RNAi technique followed by using *GFP-CENH3-tailswap* plants to rescue the CENH3 silenced cells. Selfing of co-transformed plants which carried both silencing and rescue constructs showed normal seed setting. However, when co-transformed lines were crossed to untransformed lines, a single haploid could be identified from a total of 140 progenies. Findings of this study provided indications for the use of engineered CENH3 in producing haploid inducers in polyploid crops.

As a matter of fact, centromere-mediated haploid induction is based on differences in effective centromere size between haploid inducers and their wild-type crossing partners. Wang et al. (2021) demonstrated a greatly simplified method based on crossing maize lines that were heterozygous for a *cenh3* null mutation. Crossing +/*cenh3* to wild-type plants in both directions yielded haploid progeny. Genome elimination was determined by the *cenh3* genotype of the gametophyte, suggesting that centromere failure is caused by CENH3 dilution during the postmeiotic cell divisions that precede gamete formation. The *cenh3* haploid inducer works as a vigorous hybrid and can be transferred to other lines in a single cross, making it versatile for a variety of applications.

This technique has also been advanced by different research groups across the globe by adopting alternative approaches based on CENH3 point mutations induced by EMS or TILLING. These mutants are not lethal but still can induce haploids. Plant researchers have introduced a new technique of base editing. Wang and Ouyang (2023) reported a new method that exploits adenine (A) substitution for guanine (G) to create a point mutation of CENH3 as a haploid induction line. This concept is referred to as cytosine base editor (CBE) technique. This method identified an L130F substitution within the histone fold domain in *Arabidopsis thaliana*. Haploid-inducing potential of homozygous L130F plants was subsequently tested by pollinating them with Col-0 which yielded 2.9% paternal haploid plants. This approach provides new standpoint in utilizing CENH3-mediated haploid induction in crops and wider offerings for plant breeders to exploit this in crop improvement. Conserved point mutations such as the one induced in L130F could be developed into a general instrument for haploid induction in a wide range of plant species. Further advancement in this approach will have an edge over the existing haploid production techniques.

4.3 Parthenogenesis-Induced Haploids

To utilize the haploid induction techniques, breeders are extremely interested in any developing techniques based on new haploidization principles. For this reason, studies on formation of doubled haploids in vivo during intra- and interspecific hybridization, parthenogenesis, and apomixis gained importance over time. Based on these fundamental phenomena, BBM-mediated induction of maternal haploids and haploid induction via salmon system in wheat have gained importance and are briefly discussed below.

4.3.1 The Salmon System in Wheat

Triticum aestivum strain Salmon is a form of common wheat having $2n = 42$ chromosomes. The Salmon system represents an innovative strategy for achieving haploid production in wheat and aims to transform an ordinary common wheat cultivar into a haploid producer. This system is based on studies on self-directed embryo growth inducing complete apomixes within parthenogenetic Salmon lines. The first attempt to use parthenogenesis in haploid induction was made in wheat in the 1990s (Tsunewaki and Mukai 1990; Matzk 1996). The system involves two major steps: (1) Transfer step: A parthenogenesis-inducing *Aegilops* cytoplasm is introduced to the wheat cultivar. (2) Chromosome addition step: Additionally, the cultivar receives the 1BL-1RS translocation chromosome of Salmon where 1BL represents the long arm of wheat chromosome 1B, while 1RS represents the short arm of rye chromosome 1R (Zargar et al. 2022). This replacement of chromosomal arms speculated the exchange of two main genes of parthenogenesis control, viz., Ptg (inducer) and Spg (suppressor). Later, researchers explored an α-tubulin polypeptide, P115.1, to be present in parthenogenetic lines from 3 days before anthesis. Kumlehn et al. (2001) demonstrated that parthenogenesis is an inherent feature of the isolated egg cell. RKD-TF genes were found in *Arabidopsis thaliana* and were found to control the transition from the gametophytic to the sporophytic pathway of development, suppressing parthenogenesis.

4.3.2 Haploidization via BBM Proteins

Baby Boom (BBM) is a transcription factor of the AP2/ERF family protein that plays an important role in initiating maternal parthenogenic embryogenesis without the need for fertilization. This can lead to haploid formation in various plants, including rice, maize, millet, and even *Arabidopsis*. Apomictic plants, like those influenced by BBM, produce seeds genetically identical to the mother plant. Researchers are keen on engineering apomixis into crop plants to propagate elite hybrids. Understanding the genes and pathways involved in apomixis, including BBM, opens up exciting possibilities for crop improvement (Conner et al. 2015; Khanday et al. 2019).

Researchers have been studying genes associated with apomixis, asexual reproduction by plant seeds found in over 300 plant species. One of the main genes responsible for parthenogenesis is the PsASGR-BBML gene, which is similar to the BBM genes found in *Brassica napus*. The gene is expressed in the ovule before fertilization and has been found in *Pennisetum squamulatum*. In rice, the most closely related gene to ASGR-BBML is OsBBM1, which is expressed in the zygote. In maize, three homologs of BBM have been shown to increase expression after fertilization. The ability to create apomictic crops depends on the ability of these genes to function in crops. The BBM protein has been shown to induce somatic embryos in *Brassica* and *Arabidopsis* seedlings with ectopic overexpression (Boutilier et al. 2002; Zargar et al. 2022).

4.4 Molecular Players Involved in Haploid Induction

The genetic architecture of haploid inducer lines has been thoroughly investigated and is reported to be fairly complex. Molecular players as quantitative trait loci and genes underlying have been identified. *MATRILINEAL/NOT LIKE DAD/ ZmPHOSPHOLIPASE-A1 (MTL/NLD/ZmPLA1)* gene was the first to be identified as a key regulator whose mutation causes haploid induction in maize. In addition to this, the indeterminate gametophyte1 (*ig1*) system used for paternal and the BBM-like and the patatin-like phospholipase essential for maternal haploid induction are also reported. The independent research leading to identification of these molecular players and their manipulation to exploit them in haploid breeding is discussed in sections ahead.

4.4.1 Identification of Genetic Determinant of Haploid Induction: Dissection of a Major QTL *qhir1*

Induction of haploid is a long known fact in maize under in vivo conditions, and this can be triggered in high frequencies with the aid of pollination with special genetic stocks known as *haploid inducers*. Although the phenomenon of Stock 6-induced haploidy was discovered 50 years ago, the genetic and biological mechanism of haploid induction (HI) was still unclear. In maize, haploids are routinely produced using crosses made with Stock 6. Academic and commercial routine users of Stock 6 have bred derivatives for elevated (7–15%) haploid induction rates (HIR), at the expense of increasing rates of kernel abortion (Xu et al. 2013; Zhao et al. 2013). To understand the genetic basis of HI, several studies have been conducted. HI has been determined to be a quantitatively inherited trait, controlled by a small number of genes, and this trait can be improved by means of selection (Lashermes and Beckert 1988). Several quantitative trait loci (QTLs) affecting the haploid induction rate have been mapped on maize reference genome B73. A sincere effort of Prigge et al. (2012) has resulted in identification of a major QTL *qhir1* responsible for haploid induction trait. Genetic studies carried out with segregating populations from non-inducer X inducer crosses located this QTL on chromosome 1.04 of maize. Further, genome-wide association study conducted using 51 inducers and 1482 non-inducer maize lines by Hu et al. (2016) delineated two sub-regions, viz., *qhir11* and *qhir12*, underlying major QTL *qhir1*.

Later, studies demonstrated another genetic determinant QTL *qhir 8*. This locus affects in vivo haploid induction in maize. The QTL qhir8 was mapped to a 789-kb region and was found to have an increasing effect on haploid induction rate (HIR) and embryo abortion (EmA), but not on segregation distortion (SD). The study also developed linkage markers for MAS (Liu et al. 2015).

4.4.2 Genome-Wide Association Method for Dissecting Genetic Basis of Haploid Induction

The molecular mechanisms underlying HI in maize has been thoroughly investigated by Hu et al. (2016) through a new approach of genome-wide association studies (GWAS). The approach was termed as *conditional haplotype extension (CHE)* test that allows detection of selective sweeps even under almost perfect confounding of population structure and trait expression. This group applied this test to identify genomic regions required for HI expression and dissected the combined support interval (50.34 Mb) of the QTL qhir1, detected in a previous study of Prigge et al. (2012), into two closely linked genomic segments that were found to be related to haploid induction expression. The first one was termed as qhir11 (0.54 Mb). This was composed of an already fine-mapped region but was not diagnostic for differentiating inducers and non-inducers of haploidy. The second segment, termed qhir12 (3.97 Mb), had a haplotype allele common to all 53 inducer lines but not found in any of the 1482 non-inducers. Further, re-sequencing data of one inducer with 14 non-inducers were compared which revealed that qhir12 region was an important one containing three candidate genes involved in DNA or amino acid binding. In contrast, no such candidates could be figured out for qhir11. Based on these important findings, Hu et al. (2016) proposed that the CHE test can be efficiently utilized in introgression breeding and different fields of genetics in order to detect selective sweeps in heterogeneous genetic backgrounds.

Some recent study by Nair et al. (2017) provides deeper insight into this by investigating the effect of the two sub-region *qhir11* and *qhir12* haplotypes on HIR. They used selfed progenies of recombinants in a huge F_2 population derived from a cross between non-inducer and tropically adapted haploid inducer line. Additionally, critical genomic regions underlying *qphir1* were thoroughly investigated to work out their association with maternal haploid induction component traits. Confirming the findings of Hu et al. (2016), they also proposed *qhir12* to be obligatory for the induction of haploidy. As the haplotype of *qhir11* was found to be present in some non-inducers, its chance to be a perfect candidate determining haploid induction was not much. Rather some putative candidate genes coding for DNA and amino acid binding proteins were identified in the *qhir12* region. Further, the effect of each sub-region of *qhir1* on haploid induction rate was thoroughly characterized. They used F_2 recombinants segregating for one of the sub-regions and fixed for the other that were identified in a cross between CML269 (non-inducer) and a tropicalized haploid inducer TAIL8. The haploid induction effects of *qhir11* and *qhir12* were further characterized by using selfed progenies of recombinants between these sub-regions followed by genotyping of F_3 plants. Plants homozygous for *qhir11* and/or *qhir12* were identified which were crossed to a *ligueless* tester to determine their haploid induction rates. The study provided strong evidences for *qhir11* being the major determinant affecting haploid induction ability. This locus also causes significant segregation distortion and kernel abortion, traits that are strongly associated with maternal haploid induction. Furthermore, qhir1 has been narrowed down to a 243-kb region based on the B73 reference genome, which paves the way for gene cloning (Dong et al. 2013) which have implications in developing new inducers efficiently using marker-assisted selection (Nair et al. 2017).

4.4.3 *MATRILINEAL*: A Sperm-Specific Phospholipase Triggers Haploid Induction in Maize

Among several approaches of haploid induction, identification of candidate genes was realized as the most important aspect in understanding the genetics behind this mechanism in the past few years. Kelliher et al. (2017) got a breaking success in this approach with the identification of *MATRILINEAL* (*MTL*) as the gene inducing haploidy. They used RWK (a Stock 6 derivative with approx. 13% haploid induction rate) and crossed it to two maize inbreds, namely, NP2460 and NP2391. This was followed by subsequent backcrossing to RWK to generate mapping populations. Elevated HIR in both inbred populations that co-segregated with the marker SM020SDQ in maize bin 1.04 was observed. The finding was consistent with the reports of Prigge et al. (2012) pointing out a QTL, *qhir1* at this location. The QTL was then fine-mapped and confined within a region of 0.57 Mb, and mining of this region revealed the presence of seven candidate genes. A key step in MATRILINEAL-based double haploid breeding technology lies in its ability to cause DNA damage resulting in chromosome fragmentation. This damage is considered to underlie chromosome elimination for haploid induction. This group demonstrated that *MTL* is a pollen-specific phospholipase that triggers maize haploid induction. This was established through a combination of tools including fine mapping, sequencing, genetic complementation, and gene editing that a frame-shift mutation in *MTL* gene is responsible for triggering haploid induction. Novel edits in *MTL* lead to an increased haploid induction rate (6.7%). Wild-type MTL protein was found to localize exclusively in the sperm cytoplasm. Sequencing of pollen RNAs identified overexpression of some pollen-specific genes during haploid induction which may be implicated in development of haploid seed (Denninger et al. 2014; Hamamura et al. 2014).

Significant homology at the sequence level has been observed for *MTL* genes across cereal crops. Owing to this conservation of *MTL* genes, this approach of haploid induction can be employed for achieving accelerate breeding in other crops. Parallel and independent studies named the same gene *Matrillinial* as *Not like Dad (NLD)* by Gilles et al. (2017) and *ZmPLA1* by Liu et al. (2017).

4.4.4 Haploid Induction via Modification of a Single Gene: *NOT LIKE DAD* (*NLD*)

Advancements in haploid induction techniques gained momentum led to many parallel discoveries on candidate gene identification for haploid induction by researchers across the globe. The gene *MTL* was identified in parallel and independent research by Gilles et al. (2017) while demonstrating the role of intact pollen-specific phospholipase in determining successful sexual reproduction and haploid induction in maize. This group designated this gene encoding patatin-like phospholipase as *NOT LIKE DAD (NLD)*. The gene was identified through fine mapping of the *qhir1* locus. *NLD* was identified as the only low copy gene with a functional annotation in the target region in the 80-kb confidence interval of PK6 (haploid inducer line) within this QTL region. In all surveyed inducer lines, *NLD* was found to carry a 4-bp insertion leading to a predicted truncated protein. They demonstrated that a frameshift mutation is believed to result in haploid induction as when complemented with wild-type NLD, it exterminated the haploid induction capacity. They concluded that an intact pollen-specific phospholipase (annotated as a *patatin-like phospholipase A*) is required for successful sexual reproduction, disruption of which results in induction of haploidy. This attribute of haploid induction can be exploited as an important tool in plant breeding. Targeted disruption of *NLD* and its ortholog genes may allow establishing powerful haploid breeding tools in numerous other crops.

The discovery of *NLD* as the gene causing in planta haploid induction in maize brings up new opportunities for increasing the rate of in planta haploid induction in maize and adapting this breeding strategy to other crop species. New tools such as TILLING or CRISPR/Cas9-mediated targeted mutagenesis offer possibilities to obtain new alleles for the *NLD* gene (Barrangou and Doudna 2016). In order to exploit this technique of haploid induction in other crops, one has to first identify functional orthologues in target crop species followed by looking for allelic and sequence variations responsible for haploid induction (Table 4.1).

4.4.5 Discovery of New Players of Haploid Induction

ZmDMP (DOMAIN OF UNKNOWN FUNCTION 679 MEMBRANE PROTEIN) is one more molecular player that plays a crucial role in haploid induction in maize. It was discovered by map-based cloning and further verified by CRISPR/Cas9-mediated knockout experiments (Zhong et al. 2019). Researchers have discovered that mutation of *ZmDMP* enhances and triggers haploid induction, leading to a two- to threefold increase in the haploid induction rate (Table 4.1). *ZmDMP* is highly expressed during the late stage of pollen development and localized to the plasma membrane.

In 2021, *ZmPLD3* was identified as one more candidate responsible for haploid induction in Maize by Li et al. (2021) *ZmPLD3* is one of the members from the phospholipase D subfamily, which triggers maternal HI in maize. The gene was identified through a reverse genetic strategy followed by validation through CRISPR/Cas9 system (Table 4.1). This group further demonstrated that the knockout mutants of non-inducer line imparted haploid induction capability. Mutations of *ZmPLD3* resulted in a haploid induction rate (HIR) similar to that of mtl/zmpla1/nld. Moreover, this mutation enhanced HIR by threefold in the presence of mtl/zmpla1/nld (from 1.19% to 4.13%). Sequence analysis of this gene also showed excellent conservation of *ZmPLD3* across cereals. This further emphasizes the

Table 4.1 Targeted mutations in key genes involved in haploid induction in different crops and model plants

Target gene	Gene function	Crop	Approach	HIR	References
Centromere histone H3 (*CENH3*)	Centromere-specific histone variant involved in chromosome elimination	Maize	CRISPR/Cas9	~20%	Wang et al. (2021)
Centromere histone H3 (*CENH3*)	Centromere-specific histone variant involved in chromosome elimination	Maize	Overexpression of native CENH3	16.3%	Meng et al. (2022)
Centromere histone H3 (*CENH3*)	Centromere-specific histone variant involved in chromosome elimination	*Arabidopsis*	TILLING	–	Karimi-Ashtiyani et al. (2015)
Centromere histone H3 (*CENH3*)	Centromere-specific histone variant involved in chromosome elimination	*Arabidopsis*	TILLING	–	Kuppu et al. (2020)
ZmPLA1	Pollen-specific phospholipase A that triggers HI	Maize	CRISPR/Cas9	4.7–11%	Dong et al. (2018)
ZmPLD3	Phospholipase D3	Maize	CRISPR/Cas9	4.13% in the presence of *mtl/zmpla1/nld* mutants	Li et al. (2021)
ZmDMP	*Domain of unknown function 679* involved in gamete fusion during double fertilization	Maize	CRISPR/Cas9	0.1–0.3%	Zhong et al. (2019)
MtDMP8 and *MtDMP9*	*Domain of unknown function 679* involved in gamete fusion during double fertilization	*Medicago truncatula*	CRISPR/Cas9	0.29–0.82%	Wang et al. (2022)

(continued)

Table 4.1 (continued)

Target gene	Gene function	Crop	Approach	HIR	References
TaMTL (double and triple knockouts)	Matrilineal	Wheat	*Agrobacterium*-mediated CRISPR system	18.9%	Liu et al. (2020b)
TaPLA-A and *TaPLA-D*	Pollen-specific phospholipase A	Wheat	CRISPR/Cas9	2–3%	Liu et al. (2020a)
OsMATL	Pollen-specific phospholipase A	Rice	TILLING+ CRISPR/Cas9	2–6%	Yao et al. (2018)
REC8, *PAIR*, *OSD1*, and *MTL*	Meiotic genes, matrilineal	Rice	CRISPR/Cas9	Reduced seed-setting to 11.5%	Wang et al. (2019b)
ZMPOD65	Pollen-specific peroxidase, as a new gene controlling HI	Maize	CRISPR/Cas9	2.2–17.2%	Jiang et al. (2022)
BoCENH3	Centromere-specific histone variant involved in chromosome elimination	*Brassica*	CRISPR/Cas9	0.52–1.14%	Han et al. (2024)
CsDMP	*Domain of unknown function 679* involved in gamete fusion during double fertilization	Cucumber	CRISPR/Cas9	0.09–0.40%	Yin et al. (2024)

possible use of these in vivo haploid inducer lines for other significant crop plant species.

It is known that ROS balance plays an important role in maintaining normal reproduction. The role of altered ROS homeostasis in chromosomal imbalance is well documented. Jiang et al. (2022) identified *ZmPOD65*, as a new gene controlling haploid induction. *ZmPOD65* encodes a sperm-specific peroxidase. Pollen ROS burst in vitro can trigger sperm DNA fragmentation and ultimately induction of haploidy. Targeted mutation in *ZmPOD65* gene through CRISPR/Cas9 (missense mutation at amino acid (aa) 73 causing a change from Met to Leu) induced haploidy in maize with HIR of approximately 7.7% (Table 4.1). This provides an alternate potential route to accelerate DH-based crop breeding in maize.

4.5 Gene Editing as a Tool to Induce Haploidy

Gene editing (GE) is a technique which introduces DNA mutations in the form of insertions and/or deletions (indels) or base substitutions in target sequences. GE comprises various techniques, such as the use of zinc finger nucleases (ZFNs), transcriptional activator-like effector nucleases (TALENs), and the most perused clustered regularly interspaced short palindromic repeat (CRISPR)/CRISPR-associated nuclease 9 (Cas9) system. CRISPR/Cas9 is a versatile technique that can be applied in crop refinement technology (Naik et al. 2022). It has been used to generate haploid inducer lines in crops, which are useful for the production of haploid plants (Wang et al. 2019a). The haploid inducer lines are created by knocking out genes that are involved in the formation of the egg or sperm cells. This results in the production of haploid embryos when the inducer line is crossed with a normal diploid plant. The haploid embryos can then be regenerated into haploid plants, which can be used for breeding purposes (Son and Park 2022; d'Hooghvorst et al. 2021). By virtue of the versatility of the technique, *CRISPR readily finds applications in almost all reals of plant* biology and crop improvement to *produce* high-quality agriculturally sustainable products. One such example is the haploid inducer (HI)-Edit or haploid inducer-mediated genome editing (IMGE) in grasses (Impens et al. 2023). This method combines in vivo haploid induction with CRISPR/Cas9 gene editing to create haploid inducer lines with a 2–5% haploid induction rate in different germplasms (Liu et al. 2023). Recent discoveries of genes responsible for haploid induction have allowed engineering this trait through gene editing in non-inducer varieties of different crops. Direct editing of gametes or haploid embryos increases GE efficiency by generating null homozygous plants following chromosome doubling (Bhowmik and Bilichak 2021). The combination of GE with doubled haploid methods could potentially increase throughput for the generation of improved cultivars in several crop species. Some of the notable examples having success via this technique are discussed here.

4.5.1 CRISPR/Cas9-Mediated Editing to Knock Out *ZmPLA1* Gene Inducing Haploidy in Maize

ZmPLA1 is a gene that encodes a putative phospholipase A. It is located on the QTL *haploid induction rate 1 (qhir1)* and is essential for haploid induction in Stock 6 inducer-derived lines of maize. An important study by Liu et al. (2017) demonstrates the role of a 4-bp insertion in *ZmPLA1* gene in haploid induction in maize. *ZmPLA1* is presumably the same gene which has been designated earlier as *MTL* and *NLD* (Kelliher et al. 2017; Gilles et al. 2017). Liu et al. (2017) characterized variation at *ZmPLA1* locus by constructing a bacterial artificial chromosome (BAC) library of CAU5 (a haploid inducer line derived from Stock 6). By screening BAC clones covering the mapping region of the QTL qhir1, an insertion of 4-bp (CGAG) in the fourth exon of the gene was identified when compared with the B73 reference genome of maize. This insertion of 4-bp caused frameshift resulting in 20 altered amino acids and a premature transcription termination. This anomaly led to the truncation of the encoded protein 29 amino acids.

Further analysis of this variation over hundreds of maize and several teosinte accessions revealed this 4-bp insertion to be unique and consistently present in all haploid inducer lines. It was proposed that the 4-bp (CGAG) insertion may lead to a weak or loss-of function allele of *ZmPLA1* gene which leads to induction of haploidy/HI phenotype.

To have a full-proof established function, *ZmPLA1* was edited via CRISPR/Cas9-mediated genome editing tool to knock down this gene and validate its function. Based on the sequence of the first exon, gRNA was designed and synthesized the sequence of multiple-guide RNAs and inserted into the plant gene expression vector pBUE411. The gene editing efficiency of the recombinant vectors was tested and verified in maize protoplasts. The recombinant vector with high knockout efficiency was selected and used to transform the receptor line. The transgenic plants showing sequence variations in the target region were self-pollinated to generate T1 generations. The plants so obtained were used as male to pollinate two commercially available maize hybrids, ZD958 and JK968, to test for HIR of the transgenic plants. It was interesting to observe progenies of ZD958 and JK968 with an HIR frequency of 1.85–3.51%. The average HIR of different mutants was observed to be approximately 2%, which is close to the HIR of Stock 6 (Coe 1959). This confirmation suggested that *ZmPLA1* knockout lines may have an effect on HIR in our tested plants that is comparable to gene mutation in the parent line Stock 6.

4.5.2 Heterozygosity Fixation and Haploid Induction in Rice

Production of high-yielding varieties by exploiting hybrid vigor in commercial hybrids is the most desired goal of plant breeders. But self-propagation of the hybrids is limited because of loss of beneficial phenotypes due to genetic segregation in subsequent generations. One way is the clonal propagation of hybrids through seeds which would enable self-propagation of F1 hybrids. Wang et al. (2019b) reported a strategy to enable clonal reproduction of F_1 rice hybrids through seeds. Attempts were made to fix the heterozygosity of the rice hybrid. This was accomplished by multiplex CRISPR/Cas9 genome editing targeting important genes, viz., *REC8*, *PAIR1*, and *OSD1*. These are the meiotic genes edited to produce clonal diploid gametes and tetraploid seeds. In addition to this, haploid induction via edited *MTL* gene was also achieved. In order to assess if the manipulated *MTL* homologous gene in rice could cause the same effect as in case of Maize, Wang et al. (2019b) demonstrated that editing the *MATRILINEAL* (*MTL*) gene (involved in fertilization) could induce formation of haploid seeds in hybrid rice. They edited the *MTL* gene in hybrid rice cultivar Chunyou 84. They obtained 11 of 32 transformed plants to carry frameshift mutations. Through CRISPR/Cas9-mediated gene editing, they obtained *mtl* mutants that showed normal vegetative growth, but the seed-setting rate was reduced to 11.5% (Table 4.1). Finally, they combined fixation of heterozygosity and haploid induction by simultaneous editing of all four genes

(*REC8*, *PAIR1*, *OSD1*, and *MTL)* in hybrid rice and obtained plants that could propagate clonally through seeds.

4.5.3 Improved Haploid Induction Efficiency in Wheat Through CRISPR Edited *Matrilineal*

Sun et al. (2022) performed comparative studies of wheat pollens from the wild-type and the CRISPR/Cas9-edited matrilineal mutant (*mMTL*). The study investigated sequential events in triggering and advancement of ROS activity by *MATRILINEAL*. They hypothesized that the mutated version of *mMTL*-triggered later-stage-specific ROS causes DNA damage that may contribute to chromosome fragmentation and hence chromosome elimination during haploid induction. Specific gene edits were made in the C terminal region of wheat MLT genes. This region is known to contain important functional domains (Kelliher et al. 2017). RNA sequencing (RNA-seq) analysis was performed to compare transcriptomes between the wild-type and matrilineal mutant (*mMTL*) anthers. This revealed disturbed expression of genes for fatty acid biosynthesis and ROS homoeostasis. Since fatty acids and their derivatives are essential building blocks for anther cuticle and pollen wall formation, GC-MS was performed to analyze the chemical composition of pollen. This effort coupled with scanning and transmission electron microscopy revealed an anomaly of pollen wall development in *mMTL* anthers. Matrilineal mutant pollens were observed to be defective in many respects when compared to wild-type pollen.

This thorough investigation proposed the presence of a second major wave of ROS in postmitosis *mMTL* pollen which ultimately led to chromosome elimination and haploid induction during embryogenesis. This work revealed mechanistic insights into the function of the *MTL* gene that may work in HI and uncovered potential novel ways to manipulate plant pollen for HI potency in double haploid plant breeding.

This study should provide novel ways to manipulate *MTL*-related components, such as the use of the *MTL* promoter to directly drive ROS genes that are specifically expressed in pollen to bestow them with HI potency.

4.5.4 One-Step Creation of CMS Lines in *Brassica* Through BoCENH3-Based Haploid Induction System

Cytoplasmic male sterility (CMS) allows us to harness hybrid vigor or heterosis, which leads to superior offspring (F1 progenies) with advantages over their parents and existing cultivars. Heterosis utilization depends on the use of CMS tools for the development of homozygous fertile lines and CMS lines (Wan et al. 2021). Although DH technology in several crops facilitates rapid generation of fertile lines, creation of CMS lines in practice requires multiple rounds of backcrossing, which is time-consuming and expensive (Gautam et al. 2023). Han et al. (2024) described a method for generating both homozygous fertile and CMS lines through the induction of paternal haploids. They attempted in-frame deletion and restored frameshift mutants of BoCENH3 in *Brassica oleracea* using the CRISPR/Cas9 system. The mutant thus obtained induced paternal haploids upon outcrossing. They successfully obtained two diploid CMS-B54 lines, which were morphologically similar to its maintainer B54 and showed typical male sterile traits. Haploid inducer lines were subsequently generated with CMS cytoplasm, which led to the development of CMS lines in a single step. This new DH technology of BoCENH3-based HI system has the potential to accelerate breeding in *Brassica*.

4.5.5 Maternal Haploid Inducer in Cucumber

The CRISPR technology was also extended to a major cucurbit crop cucumber. Yin et al. (2024) identified one of the genes designated as CsDMP to be involved in maternal haploid induction in cucumber through in silico tools, heterologous fusion protein expression, and gene expression. The study found a higher expression of CsDMP in cucumber male buds and mature pollen, suggesting its potential role in gametophyte development. To test this hypothesis, the researchers mutated CsDMP using the pKSE402 vector (a CRISPR/Cas9 toolkit), resulting in two homozygous csdmp mutants. These mutants showed lower pollen viability, reduced germination rate, reduced filled seed number, and increased aborted seed percentage. Further, two types of homozygous csdmp mutants were generated in the Chinese local cultivar "Xintaimici." These mutants showed lower pollen viability, reduced pollen germination rate, reduced filled seed number, and increased aborted seed percentage. The study also tested whether csdmp mutants induce maternal haploid production in cucumber. The average haploid induction rate ranged from 0.09% to 0.40%.

4.6 Conclusions and Prospects

Haploid plants and haploid-derived homozygous lines are useful in several domains of basic and applied research. DH technology offers a fast-forward mode of truly homozygous line development to expedite crop breeding programs. DH plants also find a wide application in QTL mapping, MAS, and genomic selection in addition to hybrid production. Improvement in the efficiency of DH technology supported by novel biotechnological tools such as precise gene editing can greatly accelerate plant breeding efforts in crops. There has been a resurgence of scientific activities in haploid production technology in the past 10 years. This has resulted in great success in this area, and we have seen a discovery of many interesting molecular players that provided deeper understanding of induced haploidy and embryo development. This success can be seen in the form of tangible outcomes such as availability of reliable tissue culture protocols and production of haploid-derived crop cultivars in a wide range of plant species. The potential in these areas has been deeply realized and has drawn interest of wider community of researchers including molecular biologists. Better and comprehensive understanding of the process of haploid induction will bring possibilities for the improvement of haploid induction techniques with wider delivery opportunity extendable to many agriculturally important crops.

References

Barrangou R, Doudna JA (2016) Applications of CRISPR technologies in research and beyond. Nat Biotechnol 34(9):933–941

Bhowmik P, Bilichak A (2021) Advances in gene editing of haploid tissues in crops. Genes (Basel) 12(9):1410

Bhowmik P, Ellison E, Polley B, Bollina V, Kulkarni M, Ghanbarnia K, Song H, Gao C, Voytas D, Kagale S (2018) Targeted mutagenesis in wheat microspores using CRISPR/Cas9. Sci Rep 8:1–10

Bilichak A, Sastry-Dent L, Sriram S, Simpson M, Samuel P, Webb S, Jiang F, Eudes F (2019) Genome editing in wheat microspores and haploid embryos mediated by delivery of ZFN proteins and cell-penetrating peptide complexes. Plant Biotechnol J 18:1307–1316

Black BE, Jansen LE, Maddox PS, Foltz DR, Desai AB, Shah JV, Cleveland DW (2007) Centromere identity maintained by nucleosomes assembled with histone H3 containing the CENP-A targeting domain. Mol Cell 25:309–322

Boutilier K, Offringa R, Sharma VK, Kieft H, Ouellet T, Zhang L, Hattori J, Liu CM, van Lammeren AA, Miki BL, Custers JB, van Lookeren Campagne MM (2002) Ectopic expression of BABY BOOM triggers a conversion from vegetative to embryonic growth. Plant Cell 14(8):1737–1749

Britt AB, Kuppu S (2016) Cenh3: an emerging player in haploid induction technology. Front Plant Sci 7:357

Coe EH (1959) A line of maize with high haploid frequency. Am Nat 93:381–382

Comai L (2014) Genome elimination: translating basic research into a future tool for plant breeding. PLoS Biol 12(6):e1001876

Conner JA, Mookkan M, Huo H et al (2015) A parthenogenesis gene of apomict origin elicits embryo formation from unfertilized eggs in a sexual plant. Proc Natl Acad Sci USA 112(36):11205–11210

d'Hooghvorst I, Ferreres I, Nogués S (2021) Anther culture and chromosome doubling in Mediterranean japonica rice. Methods Mol Biol 2287:333–341

Denninger P, Bleckmann A, Lausser A, Vogler F, Ott T, Ehrhardt DW, Frommer WB, Sprunck S, Dresselhaus T, Grossmann G (2014) Male-female communication triggers calcium signatures during fertilization in Arabidopsis. Nat Commun 5:4645

Dong X, Xu X, Miao J et al (2013) Fine mapping of qhir1 influ-encing in vivo haploid induction in maize. Theor Appl Genet 126:1713–1172

Dong L, Li L, Liu C, Liu C, Geng S, Li X, Huang C, Mao L, Chen S, Xie C (2018) Genome editing and double-fluorescence proteins enable robust maternal haploid induction and identification in maize. Mol Plant 11(9):1214–1217

Gautam R, Shukla P, Kirti PB (2023) Male sterility in plants: an overview of advancements from natural CMS to genetically manipulated systems for hybrid seed production. Theor Appl Genet 136(9):195

Gilles LM, Khaled A, Laffaire JB, Chaignon S, Gendrot G, Laplaige J, Bergès H, Beydon G, Bayle V, Barret P, Comadran J, Martinant JP, Rogowsky PM, Widiez T (2017) Loss of pollen-specific phospholipase NOT LIKE DAD triggers gynogenesis in maize. EMBO J 36:707–717

Hamamura Y, Nishimaki M, Takeuchi H, Geitmann A, Kurihara D, Higashiyama T (2014) Live imaging of calcium spikes during double fertilization in *Arabidopsis*. Nat Commun 5:4722

Han F, Zhang X, Liu Y, Liu Y, Zhao H, Li Z (2024) One-step creation of CMS lines using a BoCENH3-based haploid induction system in Brassica crop. Nat Plants 18:581. https://doi.org/10.1038/s41477-024-01643-w

Hu H, Schrag TA, Peis R et al (2016) The genetic basis of haploid induction in maize identified with a novel Genome-Wide Association Method. Genetics 202:1267–1276

Impens L, Lorenzo CD, Vandeputte W, Wytynck P, Debray K, Haeghebaert J, Herwegh D, Jacobs TB, Ruttink T, Nelissen H, Inzé D, Pauwels L (2023) Combining multiplex gene editing and doubled haploid technology in maize. New Phytol 239(4):1521–1532

Ishii T, Sunamura N, Matsumoto A, Eltayeb AE, Tsujimoto H (2015) Preferential recruitment of the maternal centromere-specific histone H3 (CENH3) in oat (Avena sativa L.)9 pearl millet (*Pennisetum glaucum* L.) hybrid embryos. Chromosom Res 23:709–718

Jacquier NMA, Gilles LM, Pyott DE, Martinant JP, Rogowsky PM, Widiez T (2020) Puzzling out plant reproduction by haploid induction for innovations in plant breeding. Nat Plants 6:610–619

Jiang C, Sun J, Li R, Yan S, Chen W, Guo L, Qin G, Wang P, Luo C, Huang W, Zhang Q, Fernie AR, Jackson D, Li X, Yan J (2022) A reactive oxygen species burst causes haploid induction in maize. Mol Plant 615(6):943–955

Karimi-Ashtiyani R, Ishii T, Niessen M, Stein N, Heckmann S, Gurushidze M, Banaei-Moghaddam AM et al (2015) Point mutation impairs centromeric CENH3 loading and induces haploid plants. Proc Natl Acad Sci 112:11211–11216

Kelliher T, Starr D, Wang W, McCuiston J, Zhong H, Nuccio ML, Martin B (2016) Maternal haploids are preferentially induced by CENH3- tailswap transgenic complementation in maize. Front Plant Sci 7:414

Kelliher T, Starr D, Richbourg L, Chintamanani S, Delzer B, Nuccio ML, Green J et al (2017) MATRILINEAL, a sperm-specific phospholipase, triggers maize haploid induction. Nature 542:105–109

Khanday I, Skinner D, Yang B, Mercier R, Sundaresan V (2019) A male-expressed rice embryogenic trigger redirected for asexual propagation through seeds. Nature 565(7737):91–95

Kolesnikova EO, Donskikh EI, Berdnikov RV (2021) Haploid biotechnology as a tool for creating a selection material for sugar beets. Vavilovskii Zhurnal Genet Selektsii 25(8):812–821

Kumlehn J, Kirik V, Czihal A, Altschmied L, Matzk F, Lörz H, Bäumlein H (2001) Parthenogenetic egg cells of wheat: cellular and molecular studies. Sex Plant Reprod 14:239–243

Kuppu S, Tan EH, Nguyen H, Rodgers A, Comai L, Chan SWL, Britt AB (2015) Point mutations in centromeric histone induce post-zygotic incompatibility and uniparental inheritance. PLoS Genet 11:e1005494

Kuppu S, Ron M, Marimuthu MPA, Li G, Huddleson G, Siddeek MH, Terry J, Buchner R, Shabek N, Comai L, Britt AB (2020) A variety of changes, including CRISPR/Cas9-mediated deletions, in CENH3 lead to haploid induction on outcrossing. Plant Biotechnol J 18:2068–2080. https://doi.org/10.1111/pbi.13365

Lashermes P, Beckert M (1988) Genetic control of maternal haploidy in maize (*Zea mays* L.) and selection of haploid inducing lines. Theor Appl Genet 76:405–410

Lermontova I, Schubert V, Fuchs J, Klatte S, Macas J, Schubert I (2006) Loading of Arabidopsis centromeric histone CENH3 occurs mainly during G2 and requires the presence of the histone fold domain. Plant Cell 18:2443–2451

Li Y, Lin Z, Yue Y et al (2021) Loss-of-function alleles of *ZmPLD3* cause haploid induction in maize. Nat Plants 7:1579–1588

Liu C, Li W, Zhong Y, Dong X, Hu H, Tian X, Wang L, Chen B, Chen C, Melchinger AE, Chen S (2015) Fine mapping of qhir8 affecting in vivo haploid induction in maize. Theor Appl Genet 128(12):2507–2515

Liu C, Li X, Meng D, Zhong Y, Chen C, Dong X, Xu X, Chen B, Li W, Li L, Tian X, Zhao H, Song W, Luo H, Zhang Q, Lai J, Jin W, Yan J, Chen S (2017) A 4-bp insertion at *ZmPLA1* encoding a putative phospholipase A generates haploid induction in maize. Mol Plant 10:520–522

Liu C, Zhong Y, Qi X, Chen M, Liu Z, Chen C, Tian X, Li J, Jiao Y, Wang D, Wang Y, Li M, Xin M, Liu W, Jin W, Chen S (2020a) Extension of the *in vivo* haploid induction system from maize to wheat. Plant Biotechnol J 18:316. https://doi.org/10.1111/pbi.13218

Liu H, Wang K, Jia Z, Gong Q, Lin Z, Du L, Pei X, Ye X (2020b) Efficient induction of haploid plants in wheat by editing of TaMTL using an optimized Agrobacterium-mediated CRISPR system. J Exp Bot 71(4):1337–1349
Liu D, Yang H, Zhang Z, Chen Q, Guo W, Rossi V, Xin M, Du J, Hu Z, Liu J et al (2023) An elite gamma-gliadin allele improves end-use quality in wheat. New Phytol 239:87–101
Maheshwari S, Tan EH, West A, Franklin FC, Comai L, Chan SW (2015) Naturally occurring differences in CENH3 affect chromosome segregation in zygotic mitosis of hybrids. PLoS Genet 11(1):e1004970. https://doi.org/10.1371/journal.pgen.1004970
Malik HS, Henikoff S (2003) Phylogenomics of the nucleosome. Nat Struct Biol 10(11):882–891. https://doi.org/10.1038/nsb996. PMID: 14583738
Matzk F (1996) The 'Salmon system' of wheat – a suitable model for apomixis research. Hereditas 125:299–301
Meng D, Luo H, Dong Z, Huang W, Liu F, Li F et al (2022) Overexpression of modified CENH3 in Maize Stock6-derived inducer lines can effectively improve maternal haploid induction rates. Front Plant Sci 13:892055. https://doi.org/10.3389/fpls.2022.892055
Naik BJ, Shimoga G, Kim SC, Manjulatha M, Subramanyam Reddy C, Palem RR, Kumar M, Kim SY, Lee SH (2022) CRISPR/Cas9 and nanotechnology pertinence in agricultural crop refinement. Front Plant Sci 13:843575
Nair SK, Molenaar W, Melchinger AE, Prasanna M, Martinez BL, Lopez LA, Chaikam V (2017) Dissection of a major QTL *qhir1* conferring maternal haploid induction ability in maize. Theor Appl Genet 130(6):1113–1122
Prigge V, Xu X, Li L, Babu R, Chen S, Atlin GN, Melchinger AE (2012) New insights into the genetics of in vivo induction of maternal haploids, the backbone of doubled haploid technology in maize. Genetics 190:781–793
Ravi M, Chan SW (2010) Haploid plants produced by centromere mediated genome elimination. Nature 464:615–618
Ravi M, Kwong PN, Menorca RM, Valencia JT, Ramahi JS, Stewart JL, Tran RK et al (2010) The rapidly evolving centromere-specific histone has stringent functional requirements in *Arabidopsis thaliana*. Genetics 186:461–471
Ravi M, Shibata F, Ramahi JS, Nagaki K, Chen C, Murata M, Chan SW (2011) Meiosis-specific loading of the entromere-specific histone CENH3 in *Arabidopsis thaliana*. PLoS Genet 7:e1002121
Ravi M, Marimuthu MPA, Tan EH, Maheshwari S, Henry IM, Marin-Rodriguez B, Urtecho G et al (2014) A haploid genetics toolbox for *Arabidopsis thaliana*. Nat Commun 5:5334
Ren J, Wu P, Tian X, Lbberstedt T, Chen S (2017) QTL mapping for haploid male fertility by a segregation distortion method and fine mapping of a key QTL qhmf4 in maize. Theor Appl Genet 130:1349–1359
Son S, Park SR (2022) Challenges facing CRISPR/Cas9-based genome editing in plants. Front Plant Sci 13:902413. https://doi.org/10.3389/fpls.2022.902413
Sullivan KF, Hechenberger M, Masri K (1994) Human CENP-A contains a histone H3 related histone fold domain that is required for targeting to the centromere. J Cell Biol 127:581–592
Sun G, Geng S, Zhang H, Jia M, Wang Z, Deng Z, Tao S, Liao R, Wang F, Kong X, Fu M, Liu S, Li A, Mao L (2022) Matrilineal empowers wheat pollen with haploid induction potency by triggering postmitosis reactive oxygen species activity. New Phytol 233(6):2405–2414
Tang Q, Wang X, Jin X, Peng J, Zhang H, Wang Y (2023) CRISPR/Cas technology revolutionizes crop breeding. Plants 12:3119. https://doi.org/10.3390/plants12173119
Tsunewaki K, Mukai Y (1990) Wheat haploids through the Salmon method. In: Bajaj YPS (ed) Wheat. Biotechnology in agriculture and forestry, vol 13. https://doi.org/10.1007/978-3-662-10933-5_25
Wan X, Wu S, Li X (2021) Breeding with dominant genic male-sterility genes to boost crop grain yield in the post-heterosis utilization era. Mol Plant 14(4):531–534
Wang S, Ouyang K (2023) Rapid creation of CENH3-mediated haploid induction lines using a cytosine base editor (CBE). Plant Biol 25(1):226–230

Wang D, Samsulrizal NH, Yan C, Allcock NS, Craigon J, Ulate BB et al (2019a) Characterization of CRISPR mutants targeting genes modulating pectin degradation in ripening tomato. Plant Physiol 179:544–557

Wang C, Liu Q, Shen Y, Hua Y, Wang J, Lin J, Wu M, Sun T, Cheng Z, Mercier R, Wang K (2019b) Clonal seeds from hybrid rice by simultaneous genome engineering of meiosis and fertilization genes. Nat Biotechnol 37(3):283–286

Wang N, Gent JI, Dawe RK (2021) Haploid induction by a maize *cenh3* null mutant. Sci Adv 7(4):eabe2299

Wang N, Xia X, Jiang T, Li L, Zhang P, Niu L, Cheng H, Wang K, Lin H (2022) In planta haploid induction by genome editing of DMP in the model legume Medicago truncatula. Plant Biotechnol 20(1):22–24

Watts A, Kumar V, Bhat SR (2016) Centromeric histone H3 protein: from basic study to plant breeding applications. J Plant Biochem Biotechnol 25:339–348

Watts A, Singh SK, Bhadouria J, Naresh V, Bishoyi AK, Geetha KA, Chamola R et al (2017) Brassica juncea lines with substituted chimeric GFPCENH3 give haploid and aneuploid progenies on crossing with other lines. Front Plant Sci 7:228806

Xu X et al (2013) Gametophytic and zygotic selection leads to segregation distortion through *in vivo* induction of a maternal haploid in maize. J Exp Bot 64:1083–1096

Yao L, Zhang Y, Liu C, Liu Y, Wang Y, Liang D, Liu J, Sahoo G, Kelliher T (2018) *OsMATL* mutation induces haploid seed formation in *indica* rice. Nat Plants 4:530–533

Yin S, Li S, Sun L, Shi K, Fan S, Liu X, Ren H (2024) Mutating the maternal haploid inducer gene CsDMP in cucumber produces haploids in planta. Plant Physiol 194(3):1282–1285

Zargar M, Zavarykina T, Voronov S, Pronina I, Bayat M (2022) The recent development in technologies for attaining doubled haploid plants in vivo. Agriculture 12:1595. https://doi.org/10.3390/agriculture12101595

Zhao X, Xu X, Xie H, Chen S, Jin W (2013) Fertilization and uniparental chromosome elimination during crosses with maize haploid inducers. Plant Physiol 163:721–731

Zhong Y, Liu C, Qi X, Jiao Y, Wang D, Wang Y, Liu Z, Chen C, Chen B, Tian X et al (2019) Mutation of ZmDMP enhances haploid induction in maize. Nat Plants 5:575–580

Rice and Maize Haploids

5

Kanushree Nandedkar, Zenu Jha, and Satish B. Verulkar

Abstract

Maize (*Zea mays*) and rice (*Oryza sativa*) hold a vital position as essential cereal crops in ensuring global food security. The conventional breeding methods, are time-consuming nature, coupled with the ever-changing market demands and the inability of the existing varieties to meet the actual consumption and production requirements. Hence, it becomes imperative to accelerate the breeding process and reduce the breeding time in order to achieve breakthroughs in rice and maize breeding techniques and methodologies. Notably, significant progress has been made in the field of haploid production in rice and maize, with a strong emphasis on the development of effective in vitro systems such as anther culture and microspore culture. These techniques facilitate the creation of haploid plants, thereby allowing the fixation of desirable traits and reducing the number of breeding cycles. Moreover, the integration of genome editing technologies with haploid induction opens up opportunities for precise manipulation of traits and the creation of novel genetic variations. The implementation of these methods in the generation of homozygous lines has greatly accelerated the development of improved varieties with the desired agronomic traits. Furthermore, the utilization of haploids in mutagenesis and gene editing enables the exploration of genetic diversity and the development of rice varieties resilient to climatic change. Both rice and maize have also greatly benefited from the integration of doubled haploid inducers, which offer a controlled and predictable approach to inducing haploids. Additionally, the utilization of haploids in wide hybridization and the creation of synthetic polyploids present avenues for expanding the genetic diversity and enhancing adaptability to evolving environmental conditions.

K. Nandedkar · Z. Jha (✉) · S. B. Verulkar
Department of Plant Molecular Biology and Biotechnology, Indira Gandhi Krishi Vishwavidyalaya, Raipur, India

Z. Jha et al. (eds.), *Doubled Haploids: Technological Advances and Role In Crop Improvement*, https://doi.org/10.1007/978-981-96-2339-6_5

Keywords

Rice · Maize · Haploids · Haploid induction · Doubled haploids

5.1 Haploids in Maize

Maize, possessing immense diversity and adaptation ability, holds the position of the third most significant staple crop following rice and wheat. The development of high-yielding cultivars in maize primarily relies upon the crucial approach of hybrid breeding. This approach entails the generation of pure inbred lines with desirable traits in a prompt manner, thereby achieving greater genetic gains. By employing the conventional pedigree method, it may require up to seven generations of self-fertilization to obtain homozygous inbred lines from the initial cross. In light of its significance as a cultivated plant and in order to safeguard future food security amidst the steadily increasing global population and climate change, maize research has witnessed notable progress and accomplishments with regard to the enhancement of novel and more suitable cultivars via the application of diverse methodologies. Among these methodologies, haploid and doubled haploid (DH) technology emerged as the most advantageous method for rapid development of new inbred lines by simultaneous fixation of all genes across the entire chromosomes within two to three generations. DH technology is supplanting the traditional approach to line development and has made substantial contributions to the swift generation of inbred lines and the subsequent development of hybrids.

The initial identification of a spontaneous haploid in maize was recorded in 1929 by Stadler and Randolph, with Chase later achieving the production of maize DH inbreds through spontaneous parthenogenesis in 1946 (Randolph 1932; Chase 1947). Furthermore, the findings of 0.1% incidence of spontaneous haploid induction in maize raised the possibility of using haploids in hybrid breeding (Chase 1951). The breeders, however, were restricted by a relatively low spontaneous haploid induction rate. However, it's interesting to note that Coe in 1959 discovered a higher induction rate (up to 2.3%) in crosses with the inbred line Stock 6 which thereafter served as the ancestor for subsequently created inducer lines around the world.

The large-scale adoption and effective implementation of DH technology in maize relies on the use of efficient, sustainable, and effective techniques for inducing haploids, identifying putative haploids, and doubling chromosomes (Chidzanga et al. 2017). The production of haploids is initially achieved through various techniques, such as an in vitro method utilizing tissue culture techniques or an in vivo approach utilizing haploid inducer (HI) lines.

5.1.1 In Vitro Haploid Production

In vitro haploid induction methods can be used to generate haploids from either male (androgenesis) or female (gynogenesis) gametophytic cells. Induction of haploids in maize using anther-microspore culture is more prevalent and preferable over in vitro gynogenesis (Zheng et al. 2003). The Chinese were the first to report successful in vitro androgenesis in maize, with a positive growth response frequency of about 1% (Genovesi and Collins 1982). Cowen et al. (1992) discovered American germplasm that was responsive to in vitro gynogenesis. Later, Tang et al. (2006) reported the possibility and efficacy of in vitro cultivation of pollinated maize ovaries for the induction of maize haploids. However, compared to in vitro androgenesis and in vivo haploid induction, the frequency of haploid induction was considerably lower making the procedure less effective; thus, in vitro gynogenesis in maize has not generally been substantially studied. Although a significant amount of research has been conducted on in vitro androgenesis in maize and attempts have been made to establish this method, it still remains less favorable due to its high genotype-dependent nature and the majority of genotypes being unresponsive to it. Moreover, for the responsive genotypes, several crucial factors such as anther stage, anther pre-treatment, and specific media components are necessary for successful in vitro androgenesis to occur (Genovesi and Collins 1982). As a result, in vivo haploid induction due to its practical feasibility is a successful strategy for maize breeding programs (Chidzanga et al. 2017).

5.1.2 In Vivo Haploid Production

The process of in vivo haploid maize development involves a series of steps which includes haploid induction, haploid identification, chromosome duplication, and seed multiplication of DH plants (D0) (Gupta et al. 2021). In vivo haploid induction occurs through intra-specific crosses between lines possessing specific genetic determinants, known as inducer lines and can be approached in two distinct ways, resulting in the production of paternal or maternal haploids, respectively (Chidzang et al. 2017). In the case of maternal haploids, the genomes exclusively originate from the seed parent plant, with haploid induction being induced by the pollinator parent. On the contrary, the induction of paternal haploids involves the pollinator serving as the genome donor, while the female functions as the inducer (Kermicle 1969). The key procedures for in vivo induction-based doubled haploid generation in maize are outlined below:

5.1.3 Haploid Induction

5.1.3.1 Paternal Haploids

To generate paternal haploids, the female parent is employed as the inducer, while the male parent serves as the donor of germplasm. Consequently, paternal haploids possess the cytoplasm from the inducer and the chromosomes derived from the donor plant (Chaikam 2012). Kermicle (1969) discovered the induction of haploidy through androgenesis is regulated by the recessive mutant gene ig1 (indeterminate gametophyte), which is accountable for embryological abnormalities, such as anucleated oospheres. The *ig* mutant located on chromosome 3 (Cowen et al. 1992; Kermicle 1994) may cause substitution or hybridization of cytoplasms, producing paternal haploids with the cytoplasm of the inducer female plant and chromosomes from the donor (male) plant. The ability of the mutant *ig* is maternally sex-limited which was confirmed by crossing males with or without the mutant *ig1* gene with females carrying the mutant *ig1* gene revealing that the mutant *ig1* of paternal origin has no effect on the incidence of haploid induction (Kermicle 1969). The ig gene can also be utilized to generate cytoplasmic male sterile lines, which allow for a quick assessment of the ability to maintain or restore male fertility and to transfer valuable cytoplasm (Pollacsek 1992). But because of the poor haploid induction rate (1% to 2%) (Kermicle 1994), dependency on the genotype of the male parent, and changes in cytoplasm composition from the donor genotype, using the ig mutant for paternal haploid induction is a less appealing strategy for in vivo haploid induction in maize (Chidzanga et al. 2017).

5.1.3.2 Maternal Haploids

To generate maternal haploids, the male parent is the haploid inducer line, while the female parent is the source germplasm or donor (Prasanna et al. 2012). The male parent induces the haploidy, resulting in maternal haploids that carry both cytoplasm and chromosomes from the donor parent. This method is widely used to obtain doubled haploid lines due to its high induction rate and practicality compared to androgenesis, as it utilizes the source genotypes as female parents. Consequently, the induction of haploidy through pollination can be conducted in isolated lots, with the detasseling of the source populations, ensuring that only the pollen originating from the maternal inducer remains in the field (Trindade 2022).

Two potential hypotheses have been proposed to describe the phenomenon of in vivo maternal haploid induction (Geiger 2009; Qiu et al. 2014; Tian et al. 2018; Chaikam et al. 2019a). The initial hypothesis postulates that one of the two sperm cells fails to undergo fusion with the egg cell, instead triggering haploid embryogenesis, while the second sperm cell successfully fuses with the central cell, resulting in the formation of a triploid endosperm. Alternatively, the second hypothesis suggests that one of the two sperm cells provided by the inducer is defective; nevertheless, it still fuses with the egg cell leading to the degeneration of the inducer chromosomes in subsequent cellular divisions and is gradually eliminated from the primordial cells, while the second sperm cell fuses with the central cell, ultimately leading to the generation of a regular triploid endosperm (Rober et al. 2005; Trindade 2022; Chidzanga et al. 2017; Khulbe and Pattanayak 2021).

5.1.4 Identification of Haploids

After haploid induction, only 8 to 10% of the developing seeds are haploid, while the remaining fraction of approximately 90% of the kernels possess diploidy, making them unfavorable for the production of doubled haploid (DH) lines (Boote et al. 2016; De La Fuente et al. 2017; Melchinger et al. 2013). During the mature stage, haploids can be distinguished from diploids based on distinct plant attributes such as shorter stature, slender and feeble stems, erect and narrow leaves, and a diminished growth rate. But for producing DH inbred lines on a large scale, a crucial prerequisite is the accurate and effective separation of haploid and diploid seeds prior to subjecting to chromosomal duplication to avoid the unnecessary allocation of resources and expenses associated with the management of diploids in the field (Chaikam et al. 2016).

There are several ways for identifying haploids, including flow cytometry, multispectral imaging, and genetic marker systems (R1-nj (Navajo), high oil, red root, transgenic markers, and herbicide resistance) (as reviewed by Chidzanga et al. 2017 and Trindade 2022).

- *R1-Navajo marker*: Haploid identification in maize is commonly achieved through the utilization of the R1-Navajo marker. This marker is a consequence of the expression of the R1-nj gene, which is found within the genome of the inducers (Chase 1952; Nanda and Chase 1966). The R1-nj gene is renowned for its widespread employment in this method. By virtue of its dominance, this gene imparts pigmentation to both the endosperm and embryo of the diploid seeds. The R1-Navajo (R1-nJ) gene is responsible for the synthesis of anthocyanin, thereby giving rise to the purple coloration observed in the endosperm and embryo of maize seeds. In this particular system, haploids can be identified by possessing white-colored embryos and purple endosperms, as they exclusively inherit genes from the source genotype. These haploid individuals can subsequently be selected for chromosome duplication, ultimately yielding DH lines. Conversely, diploids are characterized by possessing purple embryos and purple endosperms, as they arise from the successful crossing of the source genotype with the inducer. Such diploid seeds should be discarded. It is also imperative to dispose of a third category of seeds, which is composed of those exhibiting inhibited expression of the R1-nj gene. This inhibition renders haploid identification unfeasible. Visual examination of each seed, however, is a laborious and time-consuming process which necessitates the employment of skilled laborers in order to ensure efficient selection. Moreover, the presence of varying levels of color intensity in the embryo marker poses a challenge to the unambiguous categorization of seeds (Dang et al. 2012). Although the R1-nj gene exhibits dominant expression, its impact can be suppressed by the presence of anthocyanin inhibitor genes in the source genotype (Prasanna et al. 2012; Chaikam et al. 2015).
- *Anthocyanin marker systems*: Two additional anthocyanin marker systems based on stem and root colors were integrated with the R1-nj maker. It has been shown that haploids can be identified using the purple/red stem and sheath color marker

conditioned by the anthocyanin regulatory gene Pl (Rober et al. 2005; Rotarenco et al. 2010). In addition to the Navajo marking on the seed, Chaikam et al. (2016) created haploid inducer lines with triple anthocyanin color markers, including the expression of anthocyanin coloring in the seedling roots and leaf sheaths. The addition of the red root marker improved the ability to identify haploids among the germinating seedlings in germplasm that demonstrated total inhibition of Navajo gene when these three color marker inducers were used to identify haploids.

- *Oil content of the seeds*: The identification of haploids can be accomplished by evaluating the oil content of the seeds after pollinating the source germplasm with an inducer possessing high oil content. This method was initially proposed by Rotarenco et al. (2007) using a non-high oil inducer which proved unreliable. Melchinger et al. (2013) put forward the notion that the inclusion of a high oil content inducer is imperative for the successful identification of haploid seeds based on their oil content. This method exhibits a significant potential for enhancing the efficiency of DH technology in maize due to its automated high-throughput screening, its applicability to all maize germplasm worldwide, and its lower rates of false discoveries and false negatives, as indicated by Melchinger et al. (2013, 2014, 2015).
- *CAUHOI*, a haploid inducer with a xenia effect due to high grain oil content, was verified by Li et al. (2009). Putative haploids were identified as seeds with an oil content of more than 48.40 g/kg.
- *Near-infrared spectrometry (NIR)*: The technique of NIR employs electromagnetic radiation with a wavelength frequency ranging from 750 to 2500 nm. By employing multivariate calibration methods, this technique enables both qualitative and quantitative evaluations in a rapid and non-destructive manner, without the need for reagents (Jones et al. 2012). For instance, in the case of haploid seeds, Gustin et al. (2020) discovered an increase in protein content, while the classification of seed oil content was achieved through the implementation of skNIR (single-kernel near-infrared reflectance spectroscopy).
- *Leaf anatomy*: The efficacy of diverse leaf anatomical traits for distinguishing haploids of maize was assessed by Choe et al. (2012), with the length of the stomatal guard cells, whereas De Souza et al. (2018), on the other hand, utilized morphoanatomical traits of the guard cells, along with the number of stomata and microgametogenesis, as parameters to identify haploid, doubled haploid, and diploid genotypes of maize.
- *Seedling traits*: Seedling traits such as radicle length, coleoptile length, and the quantity of seminal roots are another method for identifying haploids. It is a quick, non-destructive, dependable, and efficient method for early haploid identification in maize. Chaikam et al. (2017) and Baleroni et al. (2021) utilized root and coleoptile length, as well as the number of lateral roots to discern between haploid and diploid seedlings in maize.
- *Transgenic herbicide resistance*: Herbicide sensitivity in donor and inducer plants to varying herbicide concentrations is exploited in the use of herbicide resistance to select haploid plants. Tseng (2012) states that for this strategy to be

effective, maize sensitivity to an existing herbicide should be mediated by a single dominant gene. Herbicide sensitivity is expected in the haploid inducer genotype, which carries the dominant gene in its homozygous condition. Meanwhile, the source genotype is tolerant of the herbicide and bears the homozygous recessive alleles for this gene. After the herbicide-tolerant source genotype and the herbicide-sensitive inducer genotype are pollinated, the resulting progeny will either be haploid seeds or heterozygous F1 seeds, which are diploid. The heterozygotes will possess the dominant sensitivity allele from the inducer, whereas the tolerant haploid plants only got a single recessive herbicide tolerance allele from the source genotype. If dominant herbicide sensitivity is present, the hybrids between the two genotypes will be heterozygous and are predicted to be more sensitive to the herbicide than the haploid plants. This indicates that haploids withstand exposure to herbicides, which kills undesirable heterozygotes. Using the herbicide Laudis, whose active ingredient is tembotrione, Tseng (2012) tested the herbicide resistance approach for haploid identification on two- to three-leaf stage maize seedlings of several genotypes (Santel 2009).

- The effectiveness of transgenic herbicide (BASTA) resistance as a reliable marker system to distinguish maternal haploids was proven by Geiger and Deimling (1994). When sensitive donors and homozygous resistant haploid inducers were crossed, maternal haploids were produced that were both susceptible to the herbicide BASTA and spontaneously doubled maternal haploids. Sexual (F1) seedlings, however, were heterozygous at the two loci and showed resistance to the herbicide.
- *Green fluorescent protein (GFP)*: GFP is another transgenic marker that has been successful in identifying haploids (Chaikam et al. 2017; Mirdita et al. 2014; Yu and Birchler 2015). By observing the transgenic-dominant GFP marker gene expressed in germinating kernels, it is possible to identify haploids in the early stages of germination in a maize haploid inducer. However, haploids will be GFP negative since the paternal GFP gene was not present during hybridization with the haploid inducer, according to Yu and Birchler (2015). Germinated diploid seeds will produce GFP fluorescence in emerging radicles and coleoptiles.
- *Flow cytometry*: Flow cytometry has also been used to identify haploids due to its precision, reliability, ease of sample preparation, and speed of result acquisition (Battistelli et al. 2013; Couto et al. 2013). It enables accurate identification of haploid seedlings among induced progeny by detecting and analyzing DNA content of cells as they flow in a sample stream using a laser beam (Dang et al. 2012; Marrone 2009). Flow cytometry has been used to assess the efficacy of the haploid-inducing approach by identifying haploids and confirming the success of chromosomal duplication (Battistelli et al. 2013).
- Dong et al. (2018) developed a clustered regularly interspaced short palindromic repeats (CRISPR/Cas9)-mediated embryo- and endosperm-specific double-fluorescence protein marker for haploid identification.

Haploid inducers possessing a combination of phenotypic markers can be potentially implemented for ensuring reliable haploid identification in a wide range of germplasm. To illustrate, employing an inducer possessing high kernel oil, R1-nj, and red root markers allows for the exclusion of the majority of diploids through the assessment of oil content, and subsequently, the putative haploids can be reconfirmed based on the presence of R1-nj and red root markers. The utilization of such inducers, equipped with multiple genetic markers, is expected to substantially enhance the efficiency and precision of haploid identification (Chaikam et al. 2019).

5.1.5 Chromosomal Doubling

Once the successful induction of haploids has been achieved, the subsequent stage would involve the cultivation of doubled haploids through the process of chromosome doubling. The presence of a single set of chromosomes in haploids hinders the production of viable gametes through meiotic cell division causing sterility. Therefore, in order to restore fertility in haploid plants, the chromosomes must undergo duplication, resulting in the creation of doubled haploids possessing two sets of chromosomes (Chidzanga et al. 2017). Chromosome duplication can be either spontaneous or can be induced artificially by the application of chemicals.

5.1.6 Spontaneous Chromosomal Doubling

Spontaneous chromosome doubling can potentially occur through various mechanisms such as somatic cell fusion, endoreduplication, endomitosis, and other unidentified pathways (Testillano et al. 2004). In maize, the frequency of spontaneous chromosome doubling is very low, unreliable, and genotype-specific (Wan et al. 1991). However, the responsiveness of haploids to chromosome duplication is a quantifiable trait, which serves as the foundation for spontaneous duplication in maize haploids. In this context, research demonstrates that the selection of individuals capable of reverting from a haploid state to a diploid state, thereby naturally restoring fertility, may support the advancement of DH lines through the integration of this trait in source populations (Trindade 2022).

Chase (1949) was the first to demonstrate spontaneous chromosomal doubling and spontaneous restoration of haploid fertility, in which haploid plants naturally generate pollen and seed. In a recent study, Molenaar et al. (2019) verified the potential for enhancing spontaneous duplication through selection, demonstrating that in two populations, the spontaneous chromosomal doubling rate could be raised from 5% to 50% with just three cycles of recurrent selection for haploid male fertility. Furthermore, this study revealed a prevalence of additive and epistatic effects, resulting in substantial improvements in male fertility of haploid plants. Moreover, as reviewed by Chaikam et al. (2019), high genetic variation and a high level of heritability, which provide support for improvement through selection, were also identified by Kleiber et al. (2012), Ren et al. (2017), Wu et al. (2017), and Ma et al. (2018).

In haploid maize, Yang et al. (2019), Ma et al. (2018), Chaikam et al. (2018), and Ren et al. (2020) have investigated genomic areas linked to spontaneous chromosomal duplication and the restoration of male fertility. These investigations demonstrate that the restoration of male fertility in maize through chromosomal duplication is an inheritable trait and thus can be mapped and integrated into the superior germplasm in maize breeding programs for production of doubled haploids (DH) (Trindade 2022).

5.1.7 Artificial Chromosomal Doubling

For chromosome duplication, antimitotic substances are necessary which prevents the formation of the spindle apparatus during cell division and allows duplication of haploid genomes (Rao and Suprasanna 1996):

- *Colchicine*, an alkaloid derived from *Colchicum autumnale* L. (autumn crocus), is most frequently used for this purpose. Prigge and Melchinger (2011) outline one of the methods that can be used for chromosomal duplication. The methodology calls for soaking seedling's hypocotyl tips (2 mm long) in a solution containing 0.06% colchicine for 8 hours at room temperature. Following which, the seedlings are washed three times and raised for 15 to 25 days in trays or pots with a substrate in a greenhouse before being transplanted into the field. Other treatment approaches, involving the infusion of colchicine to seeds (Gayen et al. 1994) and the injection of colchicine into the shoot apical meristem (Zabirova et al. 1996), have also been explained in addition to the seedling immersion method for in vivo DH generation.
- Colchicine has been investigated and used for chromosome duplication; however due to its high level of toxicity and expensive price as a reagent, various substitutes have been discovered which cause duplication of chromosomes by blocking the mitotic process.
- Wan et al. (1991) investigated the potential of four *antimicrotubule herbicides*, namely, amiprophosmethyl (APM), pronamide, oryzalin, and trifluralin, for inducing chromosome doubling in anther-derived, haploid maize callus. The findings from this investigation revealed that all four herbicides possessed the ability to induce chromosome doubling. However, it was observed that APM and pronamide exhibited greater effectiveness in inducing chromosome duplication at significantly lower concentrations. The potential of antimicrotubule herbicides for in vivo DH production was studied by Melchinger et al. (2016).
- *Nitrous oxide gas* has been experimentally evaluated as a substitute for colchicine. Kato and Geiger (2002) developed a method where maize haploid seedlings were exposed to nitrous oxide gas for 2 days at 600 kPa (Molenaar et al. (2018) suggested exposure time to be 3 days). The findings revealed that treatment during the six-leaf stage (the stage of floral primordia formation) significantly augmented the presence of fertile sectors. Despite its effectiveness, this approach has not been widely adopted due to the requirement of large equipments, such as secure gas chambers (Weber 2014; Kato 2002; Molenaar et al. 2018), and the

difficulty of applying nitrous oxide gas in vivo to mature maize plants, which is not easily feasible on a large scale (Melchinger et al. 2016).
- The chromosome doubling agent cycloalkane, which is said to have a low toxicity for mammals, low mortality rate, and a higher chromosome doubling rate in plants, has also been used to create doubled haploid maize plants and plant components (Cui et al. 2013).

5.1.8 Seed Multiplication of DH Plants

The first generation following chromosomal duplication is known as D0 generation. Under field conditions, the D0 seedlings treated with mitotic inhibitors yield D1 seeds, which represent the totally homozygous line. From source material, 5% of the haploid seeds develop into DH lines. This rate of DH conversion is influenced by a number of variables, including genetic background, precision of haploid identification system, chromosomal doubling method, and agronomic procedures in greenhouse and field environments (Gupta et al. 2021).

5.1.9 Applications of DH in Maize

A lot of focus has been placed recently on cost-effectiveness and breeding cycle shortening. Rapid production of homozygous and homogenous progenies can be achieved by the use of the DH technique. It solves the issue of the gradual decrease in heterozygosity seen in early segregating generations, making it better than traditional breeding. Given their homozygous nature, DH progenies make excellent candidates for testing and selection in a variety of environmental settings. Gupta et al. (2021) and Nathanaël et al. (2021) reviewed the following applications of DH approach in maize:

1. *In genetic mapping studies*: Correlation between genotype and phenotype is the cornerstone of applied genetic research. Both beneficial and harmful alleles can be quickly identified in DHs by accurately duplicating unique alleles (Chang and Coe 2009). DH lines are therefore especially effective in a wide range of genetic approaches, including the identification of genes responsible for a given character, gene stacking, marker identification, quantitative trait loci (QTL) analysis, linkage disequilibrium studies, epistasis studies, and the creation of genetic maps, due to the simple and accurate fixation of genomes (Liu et al. 2016, and Forster and Thomas 2010). The identification of genomic regions that affect drought tolerance (Tsonev et al. 2009), fodder quality, and resistance to the European corn borer (Foiada et al. 2015) are a few examples of how DH is used for QTL analysis.

   There are several populations that can be used for QTL mapping, including F2, backcross generations, RILS, and DH. DH is a homozygous population that can be obtained more quickly than other populations. It has an advantage over F2 and backcross populations since it can reproduce data over different seasons and years. Consequently, DH is regarded as the best mapping population to use when creating genetic maps in a variety of crops (Liu et al. 2016; Wani et al. 2018; Kumar et al. 2019; Choudhary et al. 2019). Major QTLs, qhir1, qhir8, qhir11, and qhir12, which can be pyramided to boost HIR in maize haploid inducers, were identified through comparative QTL analyses of maize HIR (Prigge et al. 2012; Liu et al. 2015; Hu et al. 2016).

2. *Use of maize DHs for breeding F1 hybrids*: The process of developing a hybrid involves two primary stages: the creation of inbred lines and the crossing of chosen parents to produce hybrid seed. When compared to traditional breeding schemes like single-seed descent (SSD), which need four to eight generations of self-crossing to produce pure inbred lines, DH technology offers a significant first-step shortcut. DH breeding is significant because it allows for the early assessment of traits on homogenous progeny. For a basic feature like disease resistance, early phenotypic selection of the original DH plants (rather than their progenies) might be performed (Nathanaël et al. 2021).
3. *Use of maize DH for cytoplasmic male sterility transfer*: DH technique could be helpful in speeding CMS-based hybrid breeding by quickly transferring the "sterilizing cytoplasm" to a genetic background of interest. In fact, backcrossing is the traditional method used to transfer CMS. Paternal haploid embryos with maternal cytoplasm are triggered by a maize haploid inducer line (based on the ig1 mutation). Thus, it offers CMS conversion in a single step and significantly streamlines CMS-based hybrid breeding (Kermicle 1969).
4. *Maize genome editing in recalcitrant lines using DH technology*: Editing a target genotype or line in maize is particularly difficult because many maize germplasms are resistant to genetic transformation and/or regeneration (Ikeuchi et al. 2019). Two teams recently used haploid inducer lines to alter resistant germplasm and get around this issue. If feasible, direct transformation or introgression from a transformable line is used to introduce the transgenic editing cassette into the haploid inducer lines. After fertilizing and directly editing elite germplasm, pollen from this transgenic haploid inducer is employed. Only about 3% of the created maternal haploid embryos in the proofs of concept were determined to have been altered (Wang et al. 2019a, b; Kelliher et al. 2019). DH plants that have undergone subsequent genome doubling are homozygous for the intended modifications.
5. RILs and/or DH populations are crossed to create *immortalized F2 (IF2) populations*, which may be used indefinitely in QTL analysis. When estimating the additive and dominant effect of QTLs, IF2 populations are considered to have an optimal genetic background and genetic information similar to F2 populations. Replication is one area in which IF2 and F2 diverge. Unlike IF2, which can replicate because each seed is representative of a single genotype, F2 cannot. These populations are helpful in maize breeding programs for analyzing the genetic foundation of grain yield and its constituents, such as kernel quality and kernel architecture (Zhang et al. 2014).

6. *Efficient utilization of landraces and wild relatives*: Maize landraces and wild relatives which possess a wide genetic base can serve as a valuable resource for enhancing the genetic makeup of modern elite cultivars. Nonetheless, the presence of lethal alleles within landraces poses a limitation to their use in crop improvement which can be overcome through the application of DH technology, which being homozygous effectively reduces the expression of lethal alleles at the haploid level and display the fixed allelic variations that exist within the diverse populations of landraces. Furthermore, DH lines that are derived from landraces exhibiting high genotypic variance, along with a rapid decline of linkage disequilibrium and the absence of population structure, can be employed for efficient association mapping (Strigens et al. 2013).
7. *DH and genome editing*: The most crucial element of a DH breeding program is a reliable haploid identification (HID) method or marker. In a DH breeding effort, a dependable haploid identification (HID) method or marker is crucial (Geiger 2009). A gene known as MATRILINEAL (MTL)/ZmPLA1 has been cloned from maize maternal haploid inducer lines in light of the R1-nj marker's inefficiency to express in specific genotypes (Kelliher et al. 2017; Liu et al. 2017), which Dong et al. (2018) later targeted in order to create haploid inducer lines using the CRISPR/Cas9 system. Likewise, in Stock 6-derived lines, TALEN-mediated deletions up to 12% increased haploid induction near the site of the 4-bp insertion (Kelliher et al. 2017). A new strategy known as IMGE has recently been developed to create genome-edited haploids using the B73 female and the CAU5 HI line bearing the ZmLG1orUB2 CRISPR/Cas9 cassette as the pollinator. Therefore, DH programs in maize can be strengthened by newly developed genome/gene editing technologies (Wang et al. 2019a, b; Kumar et al. 2020).

5.2 Haploids in Rice

Rice, a self-fertilizing crop that matures quickly within a span of 3–4 months, typically allows for two cultivation cycles per year. Numerous varieties have been developed in this crop using traditional breeding methods. However, the process of conventional breeding requires a minimum of 6 years to introduce a new cultivar, as inbreeding must be repeated over multiple years to attain purebred lines from an initial cross. Furthermore, the genetic enhancement for seed yield is seldom achieved due to the already reached limit of yield potential. Asian nations, with their high population growth rates, necessitate increased rice production to ensure food security. The occurrence of frequent disease outbreaks, pests, and insects, especially in regions with intensive cultivation, demands immediate attention. Additionally, there is a continuous shift in consumer preferences due to lifestyle changes. To address these challenges, there is a necessity to reorient rice breeding programs in order to expedite the breeding process. The production of double haploids from superior crosses can be a viable approach to streamline traditional breeding methods as it shortens the breeding cycle by immediate fixation of homozygosity and offers high-selection efficiency, creating genetic variability through the production of gametoclonal variants suitable for breeding (Devaux and Pickering 2005). Blakeslee (1922) laid the groundwork of in vitro synthesis for haploid synthesis by observing natural haploids in Jimson weed (*Datura stramonium*). Guha and Maheshwari (1964) then used anther culture in *Datura innoxia* to create artificial haploids. Since this original discovery, haploids have been found in over 250 crop species using the anther culture approach (Mishra et al. 2015). Niizeki and Oono (1968) generated the first haploid rice using MS (Murashige and Skoog) medium combined with naphthalene acetic acid (NAA).

5.2.1 In Vitro Techniques

5.2.1.1 Androgenesis

While the induction of androgenic haploids can be achieved from cultured anthers or isolated microspores in vitro, the preferred method for cereals, such as rice, is anther culture due to its convenience. In case of rice, the stimulation of anthers often results in the formation of callus initially. Consequently, the regeneration of haploids or doubled haploids takes place through a callus phase instead of direct organogenesis.

Despite being acknowledged as a useful tool in plant breeding projects, anther culture's practical applicability is limited by a number of challenges, such as its high genotypic reliance, poor callus induction and plant regeneration frequencies, and high frequency of haploid plants (Medhabati et al. 2014). The following endogenous and exogenous parameters, as reviewed by Forster et al. (2007), Mishra et al. (2016), Mayakaduwa and Silva (2023), Patra et al. (2019), Tripathy et al. (2019a), and Tripathy et al. (2019b), determine the rate of success of anther culture in rice.

Genotype

Genotype is a significant endogenous factor that influences the in vitro androgenic response among various components outlined so far. There has been a significant variation found in the androgenic response between the two primary subspecies of Asian cultivated rice, namely, *japonica* and *indica*. When it comes to rice, *japonica* varieties are more popular and receptive to microspore embryogenesis than *indica* types. According to Silva (2010), the low androgenic response of *indica* rice can be attributed to its low callusing and regeneration abilities, as well as its large proportion of albino plants. Microspore callusing and green-shoot regeneration have been found to vary significantly even within *indica* cultivars (Guha-Mukherjee 1973; Lentini et al. 1995; Tran and Vuong 2002; Rout et al. 2016; Naik et al. 2017). Also *Oryza glaberrima* has been reported to have greater callusing and regeneration potential compared to *O. sativa* according to Gueye and Ndir (2010).

Generally, genotypes that exhibit enhanced responsiveness to callus induction typically experience failure during morphogenesis, resulting from the low heritability associated with genes governing regeneration. Consequently, it becomes imperative for a successful anther culture system to ensure equal development of these two phenomena (He et al. 2006). Research conducted on the heritability of androgenic traits has indicated that more than 90% of the observed variability can be attributed to genetic effects (He et al. 2006). As a result, intersubspecific hybridization has shown a promising potential in alleviating the genetic recalcitrance of *indica* rice to a certain degree (Zhang 1989; Herath and Bandara 2011). Notably, F1 hybrids originating from crosses between *indica* and *japonica* varieties have demonstrated superior outcomes in terms of callus induction and plant regeneration compared to pure indica parents (Yan et al. 1996).

Donor Plant Environment

The outcome of the anther culture is strongly influenced by the growth conditions of the plants that produce anthers. Accordingly, alterations in the environment in which plants are cultivated may prevent even the most responsive genotypes from succeeding (Foroughi-Wehr and Wenzel 1993). It has been reported that the frequency of embryogenic pollen grains (P-grains) is directly influenced by the conditions under which the donor plants are grown (Heberle-Bors 1985). The varying temperatures during the booting stage of the donor plants have been shown to affect the normal development of microspores in vivo as well as the androgenesis of microspores in vitro (Raina 1989). The optimal level of nitrogen in the donor plants leads to anthers with superior microspores that have high embryogenic ability (Lapitan and Violeta 1999; Prem et al. 2004). However, it is not desirable for plants to have excessive vegetative growth, and therefore, the second application or foliar spray of nitrogen is avoided (Prem et al. 2004).

Physiology of the Donor Plant

The physiology of the donor plant is a crucial contributing factor for the successful cultivation of rice anthers. Although the specific growth conditions of the donor plants have an impact on the repeatability of anther culture success, it has been observed that cultivating plants under strictly controlled environments does not yield the optimal results. In the case of rice, anthers obtained from field-grown plants have shown a superior response during in vitro culture compared to anthers harvested from plants grown under controlled conditions, such as greenhouses or pots (Veeraraghavan 2007). Furthermore, it has been noted that the age of the donor plant is strongly correlated with its physiological processes, with anthers from first-formed flowers or the beginning of the flowering period exhibiting greater competence during in vitro culture (Silva 2012; Smykal 2000), while this competence decreases with the age of the plants (Szarejko 2003). Anthers derived from the primary tillers have also been found to be more responsive than those from secondary tillers (Asif 2013). Additionally, anthers obtained from the middle portion of the panicles in the boot stage have been observed to be favorable for callusing and regeneration (Jacquard et al. 2006).

Developmental Stage of Anther/Microspore Maturity

The microspore stage holds significant importance in relation to the androgenic response. According to Datta and Wenzel (1998), the early to mid-uninucleate pollen stage in rice is the most favorable. Older pollens, which are at the tetrad stage and after the first pollen mitosis, do not respond to culture because of starch deposition, which causes differentiation into male gametophytes (Cha-um et al. 2009; Silva 2010). A discernible morphological trait of the plant, which demonstrates a strong correlation with the stage of pollen development, can be employed as a means to determine the necessary microspore stage (Nurhasanah et al. 2016). In general, anther maturity can be accurately determined by measuring the distance between the flag leaf's collar and the ligule of the tiller's penultimate leaf (Bishnoi et al. 2000).

Pre-treatments

Pre-treatment for embryogenic induction of microspores can be applied physically or chemically to whole or excised inflorescences, excised anthers, or even anther donor plants (Touraev et al. 1997). During the embryonic period of pollen grains, stress applications such as temperature pre-treatment, osmotic shock, and sugar starvation are critical for inducing androgenesis (Shariatpanahi et al. 2006). The type, duration, and timing of these pre-treatments, however, might vary based on the species or even variety of rice (Datta 2001).

Cold Pre-treatment

Low-temperature pre-treatment is one of the several stimuli that encourage androgenesis and is frequently used in rice because of its effectiveness. According to Kiviharju and Pehu (1998), the benefits of cold pre-treatment on callus induction include the delay of anther wall senescence, an increase in pollen grain symmetry, and the release of chemicals essential for androgenesis, primarily shock-thermic proteins and amino acids. Herath et al. (2009) found that cold pre-treatment at 8 °C for 14 days is the most beneficial for anther culture in selected *indica* and *japonica* rice varieties and intersubspecific hybrids. Mishra et al. (2013) found that elite *indica* rice hybrids benefited from a pre-treatment at 10 °C for 7–9 days. A recent study by Kaushal et al. (2014) revealed that cold pre-treatment for 5 days at 12°C yielded the most promising results for callus induction and plant regeneration in 13 *indica* rice varieties.

Osmotic Stress

The disruption of plasmodesmatal connections between pre-embryonic cells caused by osmotic stress or loss of cellular water content results in physiological isolation of the cells, allowing a higher number of cells to differentiate and grow independently into adventive embryos (Wetherell 1984). According to several research, green plant regeneration frequency is increased from 3 to 33% when mannitol (0.4 M) is used to induce osmotic stress in microspore culture in *indica* (IR 43) and *japonica* types (Cistue et al. 1994; Pande 1997; Raina and Irfan 1998). In *indica* rice cultivars IR43 and Pusa Basmati 1, an increase in agarose concentration from 0.5% to 1.0% results in a shoot regeneration increase of over eight times (Silva and Ratnayake 2009).

Nutrient Starvation

It has been noted that under food shortage, pollen grains undergo cytoplasmic and nuclear alterations which includes dedifferentiation of plastids, the presence of a large vacuole, the loss of nuclear pores in the vegetative nucleus, structural alterations in the chromatin and nucleolus, and nucleolus size reduction (Garrido et al. 1995). P-grains with inherent induction potential were efficiently formed when pollen was starved, either by eliminating sugar or nitrogen from the in vitro environment (Heberle-Bors 1989). According to Ogawa et al. (1995), sugar starvation of the IR43 anthers for 2 days at the start of culture promotes androgenesis (39%, 12-fold), which is more than the effects of the other mannitol pre-treatments.

Culture Media

Culture medium set the direction of growth in addition to providing nutrients to the tissues. The two most popular media for rice anther cultivation are N6 (Chu 1978) and MS (Murashige and Skoog 1962). Nevertheless, by altering the nitrogen rates and sources, carbon content and sources, and vitamin concentrations and rates, a number of additional medium formulations inducing N6-like Chaleffs R-2, Heh-5, MSN, SK1, SK-3, SK-8, RZ, Szechuan medium, Medium V, and Heh-2 had been developed by 1978 which were found to be encouraging for anther culture response in rice. Carbohydrates, owing to their nutritional and osmotic properties, are necessary for androgenesis. In the induction media for rice anther production, sucrose (2–5%) has been employed as the main source of carbohydrates (Reinert and Bajaj 1977). However, when sucrose content in the induction medium exceeds 6%, it frequently increases the proportion of albino plants (Wang et al. 1978). Thus, maltose is better than sucrose as a carbon source in rice anther culture for callusing, as indicated by Naik et al. (2017) and Sen et al. (2011); additionally, maltose was found to reduce the frequency of albino plants from anthers when compared to sucrose (Park et al. 2013).

Another factor influencing androgenesis is nitrogen content of culture media that is given in the form of nitrate and/or ammonium ions. Ammonium nitrogen (NH4+) is usually needed in small quantities for cereal anther production. According to Tripathy et al. (2019b), the N6 medium, which is low in (NH4+) and high in NO3−, has shown to be highly effective for rice anther culture. Furthermore, Herath et al. (2007) demonstrated that increasing the rate of KNO_3 nitrogen and reducing the rate of NH4+ in three distinct mediums (N6, B5, and Miller) enhanced the frequency of callus induction.

Furthermore, in vitro cultures of different tissues have employed amino acids as a source of nitrogen to improve the regeneration and growth of green plants (Ogawa et al. 1995). According to studies by Faruq et al. (2014) and Sah (2008), callus induction and plant regeneration in *indica* rice are enhanced by the addition of $AgNO_3$, which is known to suppress ethylene responses in anther culture Zhao et al. (2002).

Plant Growth Regulators/Hormonal Requirements

Maintaining an appropriate mix of plant growth components in the culture media is just as crucial when dealing with genotypes that may require differently designed nutritional regimens throughout different stages of in vitro growth (Ali et al. 2021). According to Trejo-Tapia et al. (2002) and Mukherjee et al. (2015), the main factors influencing embryogenic callusing in rice anthers appear to be 2,4-D and NAA either individually or in combination with kinetin in the culture media. Regeneration cannot occur in the presence of 2,4-D or NAA alone in the media; instead, cytokinins such as Kn and BAP in combination with NAA are necessary (Mandal and Gupta 1995).

Culture Conditions

Androgenic success can also be influenced by the conditions in which tissues are incubated. Optimizing the content and circumstances of the culture is crucial to ensuring the enhancement of both the induction potential and embryogenic potential of induced calluses. Regarding the impact of light quality on anther culture, Bjørnstad et al. (1989) observed that, though low-intensity white light or darkness are considered encouraging, high-intensity white light inhibits the embryogenic induction of microspores. It has been discovered to be crucial for anthers to be constantly incubated in darkness.

Gelling Agent

Agar is the most frequently used gelling agent for solidifying media because of its inexpensive nature and wide availability; nonetheless, the release of contaminants from agar into the medium hinders growth and development. To avoid this issue, different gelling agents might be used in place of agar (Mishra et al. 2016). Alternatively, ficoll (a sucrose-based synthetic polymer) could be employed to increase surface density and improve the ratio of green to albino plantlets (Lashermes 1992), to keep embryoids from sinking in anther culture (Kao 1981), and to maintain optimal osmotic pressure. Gel rite, an agar-free gelling agent, appears to be the most effective, according to Lee and Lee (1995) (Tripathy et al. 2019b).

Interactions between genotypes, culture medium, and culture temperatures are also critical in activating resistant genes for callus induction and regeneration (Javed et al. 2012). Ratheika and Silva (2007), Talebi et al. (2007), Niroula and Bimb (2009), and Silva (2010) demonstrated that changing medium compositions (high doses of nitrogen, phosphorus, and potassium) could improve anther culturability; also, their findings suggested selection of more specific rice genotypes along with manipulating non-genetic parameters such as culture medium components and pre- and post-culture conditions can improve rice anther culturability.

5.2.1.2 Gynogenesis

The early 1980s witnessed the achievement of haploid plant production using megaspore cells through the independent research efforts of various laboratories (Kuo 1982; Yang and Zhou 1982; Zhou and Yang 1980, 1991). This approach has proven a useful substitution, particularly for rice cultivars that are androgenically resistant and produce albinos (Bhat and Murthy 2007; Chen et al. 2011; Chen and Lin 1981). The success of megasporogenesis, however, relies on the transition from the late uninucleate to the early binucleate stage of the ovary. Remarkably, the advantageous results were obtained when unhusked flowers with receptacles affixed to pistils and stamens were employed (Gosal et al. 1997). Megaspore-derived proembryos have been shown to produce callus, which may lead to the regeneration of albino plantlets (Davoyan 1985; Keller and Korzun 1996; Zhou and Yang 1981). But the laborious process and the fact that each ovary contains only one embryo sac as opposed to hundreds of microspore cells per anther which reduces the likelihood of successfully inducing haploids, restricts its advantage over anther culture (Gosal et al. 1997; Zhou and Yang 1980; Kasha and Maluszynski 2003; Murovec and Bohanec 2012). With the exception of a few crops like sugar beet and onions, gynogenesis is not widely employed in commercial applications (Kasha and Maluszynski 2003; Murovec and Bohanec 2012).

5.2.1.3 *Indica* Versus *Japonica* Rice

The *japonica* and *indica* subspecies of rice have long been acknowledged as the two main Asian cultivated varieties. According to Yang et al. (2014), *japonica* rice which is primarily grown in more temperate climates at higher latitudes or altitudes is sticky in nature, whereas *indica* rice which is primarily grown in tropical and subtropical conditions at lower latitudes or altitudes is not sticky. While anther culture is simple to implement in *japonica* rice, it is more difficult to breed through in *indica* rice limiting its use. Because of early anther necrosis, poor callus proliferation, and regeneration of albino plantlets, *indica* rice cultivars show low anther culturability (1.2% callus induction) as compared to *japonica* cultivars having 20-fold higher (28.1%) anther culturability (Grewal et al. 2011) and, thus, resulting in low recovery of doubled haploids and making androgenic applications less effective for *indica* rice (Senadhira et al. 2002; Chen et al. 1991).

Using efficient techniques to increase the ratio of green to albino plant regeneration (Samantaray et al. 2021), transforming haploids into doubled haploids by using antimicrotubular agents, and removing unwanted heterozygous diploids (Ahmadi and Ebrahimzadeh 2020) are also necessary for the successful adoption of androgenesis in *indica* rice. The emergence of the salt-tolerant variety "PSBRc50 Bicol" from a cross between two *indica* breeding lines (IR 5657-33-2 _ IR 4630-22-2-5-1-3) is one of the success stories in *indica* rice anther culture. Although there are a few unique instances of effective anther culture-based novel *indica* variety generation (Mohiuddin et al. 2014; Naik et al. 2017), the method is not routinely employed for *indica* rice breeding.

5.2.1.4 Challenges of In Vitro Methods

While the in vitro techniques are acknowledged as useful tools in rice breeding, they also have a number of challenges (Fig. 5.1), including the need for technical proficiency, cultural standardization, and genotype dependency, which makes them expensive, time-consuming, and labor-intensive. More precisely, its priority for applied research is hampered by low rates of chromosomal doubling, albinos,

Fig. 5.1 Steps for *in vitro* haploid production in rice

induced alterations, and plant regeneration (Dunwell 2010; Jacquier et al. 2020; Yao et al. 2018). The donor plant can only be used during a certain season due to particular growth conditions, which eventually prohibits its large-scale utilization (Qian et al. 2000).

5.2.1.5 In Vivo Methods of Haploid Production

The challenges of in vitro-based approaches led breeders to shift to new HI technologies. The in vivo approaches involve non-tissue culture-based procedures for inducing haploidy such as wide hybridization, CENH3-based haploidization, and use of HIS which are discussed below:

5.2.1.6 Wide Hybridization

In cereal crops, wide hybridization is used as a way to produce haploid plants in addition to in vitro techniques like anther and microspore culture for haploid induction. According to Dwivedi et al. (2015) and Watts et al. (2018), this technique crosses two divergent genomes to create a single nucleus, which is then followed by selective genome removal to produce haploid embryos. Because the combined genomes are incompatible, chromosomes from one genome are frequently eliminated from the developing embryo in certain wide crosses even while fertilization between gametes of two separate species produces a healthy embryo (Mayakaduwa and Silva 2023). The uniparental chromosome removal in the wide hybridization technique has been explained by a number of theories. Among them are (1) parent-specific centromere inactivation (Finch 1983), (2) non-synchronous nucleoprotein synthesis (Bennett et al. 1976), (3) host-specific nucleases that degrade chromosomes (Davies 1974), and (4) differences in cell cycle timings between the parents (Sanei et al. 2011). It is a robust approach that can produce haploid embryos in up to 30% of the pollinated florets; however its effectiveness varies with genetic factors and environmental conditions (chromosome elimination takes place more efficiently at temperatures over 18 °C (Kumlehn 2014).

Using the principle of wide hybridization, the "bulbosum" technique was the first approach for producing haploids in cereals proposed by (Kasha and Kao 1970). In this technique, a cross between *Hordeum vulgare* and *H. bulbosum* results in the elimination of the chromosome of *H. bulbosum* during embryonic development and thus leaving most seedlings with only one pair of *H. vulgare* chromosomes. Although the approach of wide hybridization has been somewhat successful in inducing haploids in crops like barley, maize, and wheat, it is not effective for rice since wide hybridization in the genus *Oryza* has not yet produced haploids.

Using rice × maize and rice × pearl millet crosses, Kyum et al. (2021) examined the potential for HI in rice; however, fertilization was not successful. According to Kumari's (2019) microscopic observations of pollen germination and pollen tube formation, prefertilization barriers may be inhibiting the HI pathway. For a possible two reasons, this method has not been used for haploid generation in this particular case of rice. It was difficult to switch from a well-established in vitro-based technology to a novel one for producing haploid plants, and the development of incompatible or sterile hybrids made it difficult to make use of in vivo HI capabilities. There have been successful crossbreeding attempts with numerous wild rice species, which showed considerable variabilities in segregating generations, and the plants that resembled haploid phenotypes were rarely found, which were rejected because the crossings were attempted under strict guidelines. Therefore, to investigate the potential of any wild species of rice for HI ability, planned efforts are needed (Kyum et al. 2021).

5.2.1.7 CENH3-Mediated Haploid Induction

Wide hybridization is a useful technique for haploid induction in a number of plant species, but the precise process of chromosome elimination was not well known until recently. The elimination of chromosomes in developing embryos of hybrid origin was previously thought to be caused by centromere incompatibility and asynchrony between cell cycles (Bennett et al. 1976; Finch 1983). However, recent investigation has not only clarified the cellular and molecular mechanisms underlying this process, but it has also sparked interest in using inter- and intra-specific hybridization as an effective way to induce haploidy in various crops.

Centromeres control the kinetochore assembly process during mitosis and meiosis and are essential for the segregation of chromosomes. Kinetochores facilitate chromosomal mobility during nuclear division by interacting with the spindle's microtubules. Among the several proteins that make up the kinetochore is a histone H3 variation called CENH3, which is known to be essential for the proper segregation of chromosomes. CENH3 replaces normal H3 in centromeric nucleosomes. Consequently, in an effort to both engineer haploid induction in plants and clarify the process of chromosome elimination after wide hybridization, this centromere-specific CENH3 has been the subject of extensive investigation (Kyum et al. 2021; Mayakaduwa and Silva 2023).

While researching the organization and functionality of the centromere, Ravi and Chan (2010) discovered that modifying the CENH3 protein results in the formation of haploid plants. This unintentional discovery gave birth to a unique way of HI. The CENH3 protein is made up of two domains: the N-terminal tail domain and the C-terminal histone fold domain (HFD). According to studies, the tail domain is widely variable, whereas the HFD is evolutionarily conserved among species; hence, triggering minute changes in the HFD may significantly impact its activity, resulting in a lethal phenotype. As a result, researchers have attempted to change the N-terminal domain using genetic engineering methods in order to produce haploids (Britt and Kuppu 2016; Jacquier et al. 2020; Lv et al. 2020; Watts et al. 2016).

For CENH3-mediated haploid plant generation, there are now two different methods available. Creating CENH3 null mutants and complementing the null mutants with transgenic constructs or replacing them with distant species are the two steps in the first strategy. Finding CENH3 mutants in the TILLING population or using genome editing mutants to induce haploids with minimally harmful effects is the second strategy (Britt and Kuppu 2016; Kalinowska et al. 2019). CENH3 protein shows hypervariability for HI from plant to plant; therefore, CENH3-mediated HI is not well characterized in crop plants except *Arabidopsis*. Nevertheless, the method can be applied to any crop species as long as naturally occurring cenh3 mutants can be found in its germplasm or generated through gene editing, as shown in the schematic diagram in Fig. 5.2.

The discovery of a solitary gene that encodes the CENH3 protein in rice, as reported by Nagaki et al. (2004), and subsequently its isolation and characterization by Hirsch et al. (2009), has provided valuable insights into the functionality of the centromere. In rice, the CENH3 nucleosomes are composed of 155-base pair CentO satellite repeats, which are interspersed with the centromeric retrotransposon of rice (CRR). Moreover, these repeats exhibit significant variation in terms of both sequence and length when compared to closely related species, as documented by Watts et al. (2016) and Zhang et al. (2013). In their investigation, Hirsch et al.

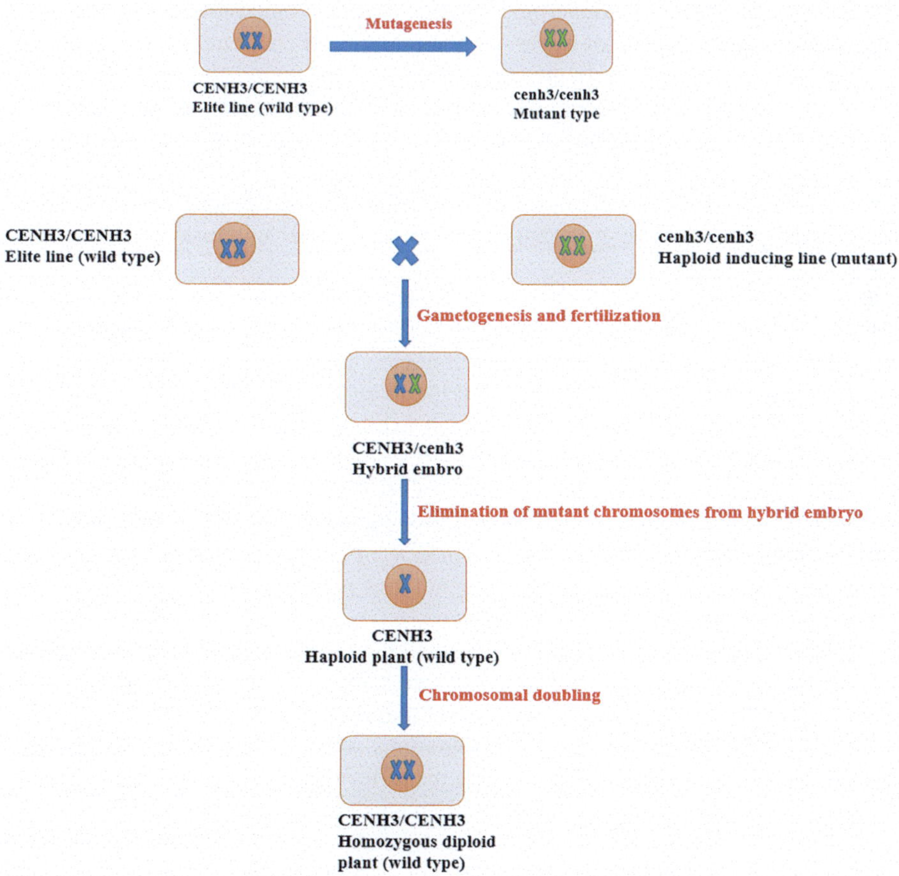

Fig. 5.2 Steps for developing CENH3 based haploid inducer system in rice

(2009) explored the expression and evolutionary patterns of CENH3 genes in rice, revealing their robust conservation from diploid to allotetraploid species. Notably, their findings demonstrated that the process of allopolyploidization did not impact the expression patterns of CENH3 genes in the *Oryza* species.

However, recently Kalinowska et al. (2019), in their review, identified a patent that describes the development of three haploid-inducing rice lines. This was achieved through the substitution of single amino acids in the N-terminal region of CENH3, resulting in the production of up to 1% haploid progeny. Additionally, a recent study in the genus *Oryza* identified and characterized several variants of CENH3 (Kaur et al. 2023). However, there is still a need for a better understanding of the consequences that arise from manipulating CENH3-linked genes (Kalinowska et al. 2019; Wang et al. 2019a, b). Furthermore, the precise mechanism of this technique in the case of rice requires further in-depth elucidation in order to achieve a significant rate of haploid induction.

5.2.1.8 Use of Haploid Inducer Lines/Stocks

The utilization of haploid inducer lines for the induction of haploids has emerged as a potent technique, capable of bypassing the majority of challenges associated with the aforementioned methods in the context of rice. This approach entails the simple way of crossing haploid inducer lines or stock with the recurrent parent, resulting in the generation of haploid embryos within the plant (Mayakaduwa and Silva 2023). Especially due to the *indica* cultivars' suitability over in vitro techniques, the in vivo creation of haploid seeds with the application of HIS has the potential to revolutionize the rice improvement program (Dwivedi et al. 2015; Saurabh 2021; Watts et al. 2018).

In an experiment, Yao et al. (2018) used CRISPR/Cas9-mediated genome editing of the terminal region of the rice ZmMTI/PLP/NLD orthologue to generate the HI traits. Through a succession of knockout mutants of the target gene, they were able to reduce the quantity of seeds while still inducing 2–6% HIR. The rice matl gene was given the OsMATL designation since it exhibited similar results to the maize matl gene. Their research demonstrated matl's potential for creating better HIS in rice and other crops. The application of the CRISPR/Cas9 technique to knock out the OsMATL gene and thereby induce in vivo haploids laid the groundwork for a new breeding pathway in rice. The frequency of haploids obtained through the knockout of the OsMATL gene in rice genotypes is nearly identical to that of the original Stock 6 of maize.

Following this, Liu et al. (2021) performed the development of haploid inducer lines in the *indica* rice line IR58025B through the utilization of CRISPR/Cas9-mediated genome editing by targeting exon 1 and exon 4 of the OsMTL gene, which resulted in the creation of a series of mutations which, when selfed, exhibited a haploid induction rate (HIR) of 6%, while the HIR ranged from 2 to 5% during outcrossing. The steps in the generation of DH plants are outlined in Fig. 5.3.

Additionally, Wang et al. (2019a, b) investigated the combined impact of OsMTL gene mutations and three other genes that are essential for governing meiotic processes. Mutations in the last three genes caused substitution of meiosis by mitosis

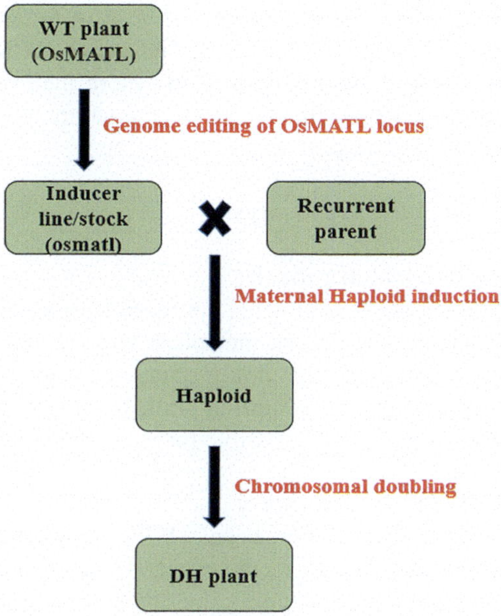

Fig. 5.3 Steps for developing OsMATL based haploid inducer system in rice

generating the MiMe (mitosis instead of meiosis) phenotype, which produces male and female clonal diploid gametes, by substituting a division that resembles mitosis for meiosis. All three of the MiMe genes were mutated in an *indica-japonica* origin hybrid rice variety, and it was found that the triple mutant F1 hybrids generate diploid clonal gametes, indicating that meiosis was converted into a mitosis-like division, which is compatible with the MiMe phenotype. However, it was required to trigger haploid induction in the transformants because self-pollination of MiMe plants doubles the ploidy level at each generation. Therefore, Wang et al. (2019a, b) simultaneously altered the OsMTL gene and showed that it could be utilized to fix heterosis in self-pollinating rice hybrids when paired with MiMe technology. This has led to the development of a novel technique for clonal seed propagation of hybrid rice. Moreover, using CRISPR/Cas9 editing of the OsMTL gene, Wang et al. (2022) successfully generated a large number of haploid-inducing rice lines with varying heading dates.

Recently, Jang et al. (2023) have successfully identified the protein OsMATL2, which functions as a phospholipase enzyme. This protein has been found to be highly expressed in the pollen of the *japonica* rice plant, particularly in the plasma membrane of its cells. To investigate the effects of OsMATL2 gene, the CRISPR/Cas9 system was employed to generate knockout mutants (designated as sgOsMATL2) of the OsMATL2 gene with suppressed expression of the OsMATL2 protein. Interestingly, in comparison to normal rice plants, diploid sgOsMATL2 plants did not exhibit any vegetative abnormalities. However, their haploid counterparts were found to be of reduced sizes. Furthermore, all sgOsMATL2 plants displayed reduced seed setting (around 80% of usual), and most significantly, serious irregularities in pollen formation were found to be the cause of male sterility in haploid sgOsMATL2 plants. The average HIR of sgOsMATL2 plants was found to be 6.34%, which is marginally more than the HIR brought on by the OsMATL mutation suggesting the promising nature of OsMATL2 gene for haploid-induction in *japonica* rice.

5.2.1.9 Ploidy Analysis

Cytological Approach

The classical cytological approach by counting the chromosomes of root tip cells that are arrested in the metaphase stage was believed to be the most effective method for ascertaining the ploidy status of plants. However, the method is laborious and time-consuming, requiring skill; moreover, tissues with dividing cells might not always be easily accessible for study (Mishra et al. 2015).

Use of Flow Cytometry

According to Martínez et al. (1994), flow cytometry made it possible to quickly and accurately determine the ploidy level of a plant generated from an anther. Flow cytometry by examining the DNA content of plants produced from microspores determines their ploidy level by comparing the results to a known diploid plant. Also as a control for the 2C DNA content, diploid donor plants are also evaluated in parallel to microspore generated plants (Ingle et al. 2023).

Morphological or Phenotypic Marker

To develop phenotypic markers in rice, Wang et al. (2023) evaluated numerous germplasm lines and identified the NWFB10 variety, which has hairy leaves at the seedling stage and can be utilized as a phenotypic marker. The dense trichomes present on the leaf surface of NWFB10 seedlings at 10 days after germination stage grow more prominently as the plant matures. To confirm that hairy leaf may be used in haploid identification, a cross between NWFB10 and 93–11 variety (hairless leaf) was performed, and it was found that all of the F1 offspring had hairy leaves. In the segregating population of 108 F2 plants, 76 plants exhibited hairy leaf phenotype identical to NWFB10, whereas 32 plants exhibited hairless leaves, fitting the typical segregation ratio of 3:1.

Molecular Markers

Simple sequence repeat (SSR) markers which are regarded as the most important option for genotyping are widely distributed across the rice genome and are highly reproducible compared to other PCR-based markers (Song et al. 2019) which can be used in determining the source of the origin of callus, i.e., whether callus came from the mother cells of microspores or from diploid somatic tissue during embryogenesis (Ingle et al. 2023). In rice hybrids, homozygous spontaneously doubled haploids have been effectively identified using SSRs (Samantaray et al. 2021).

5.2.1.10 Applications

The various applications of haploids and double haploids in rice as reviewed by Mishra and Rao (2016) are discussed below:

1. *Shortening of breeding cycle*: In order to develop new varieties, quickly and effectively (Tuvesson et al. 2007), haploid system offers an unmatched opportunity to shorten the breeding cycle and fix agronomic traits in the homozygous state, such as recessive genes for disease resistance (Datta 2005).
2. *Genetic mapping*: Haploids or doubled haploid (DH) lines are highly advantageous for genetic mapping due to their inherent homozygosity and uniformity. Furthermore, each DH line has the capability to be replicated indefinitely across multiple locations. In the context of rice, the development of a linkage map using a DH population has proven instrumental in the identification of molecular markers that are closely associated with major genes conferring resistance to rice blast, bacterial blight, and sheath blight disease (Wang et al. 2001). The amalgamation of genetic and physical maps within the DH population has furnished the necessary precision in pinpointing potential candidate resistance genes against these diseases (Wang et al. 2001). Similarly, the utilization of DHs has facilitated the establishment of chromosome maps in various species, including rice, rapeseed, and wheat (Forster and Thomas 2005).
3. *Mutational studies*: A DH system plays a crucial role in the induction and fixation of mutations. It serves to enhance the efficiency of mutation induction and selection and is instrumental in stabilizing and expressing desirable recessive traits that have been introduced through mutation or hybridization, thereby enriching the germplasm. Rice varieties treated with ethyl methanesufonate (EMS) at 10 and 20 days after anther inoculation have exhibited a high frequency of rice mutants (Lee and Lee 2002). Stable mutants with significantly higher and earlier flowering than their parent have been observed in gamma-irradiated mutagenic lines when subjected to anther culture (Myint et al. 2005).
4. *Marker-assisted selection studies*: The efficacy of the bulked segregant assay heavily relies on precise phenotyping. The DH population possesses a distinct advantage in this regard due to its homozygous or pure line characteristics, which are frequently employed in the practice of bulked segregant analysis, a widely utilized technique in marker-assisted breeding (William et al. 2002).
5. *Enhancing grain and nutritional quality*: The primary limitations that hinder the widespread adoption of hybrid rice cultivars are the cost of seed production and the poor quality of the grain in the hybrids. The technique of anther culture presents a significant opportunity for expediting breeding advancements and enhancing the characteristics related to grain quality (Xa and Lang 2011). Furthermore, plant breeders are actively striving to improve the nutritional quality of the grain. These challenges can be overcome by augmenting the micronutrient content of the existing and future high-yielding indica varieties through the utilization of donors with high levels of zinc and iron found in *japonica* rice, which have been proven to respond well to anther culture. To address this, IRRI has embarked on an ambitious project, supported by the HarvestPlus Challenge Program, with the

aim of producing nutritious rice. The generation of DH lines is a crucial component of this mission, and over 1500 DH lines have been assessed for their agronomic potential and their levels of iron and zinc content through anther culture evaluation (Grewal 2009).

6. *QTL mapping*: QTLs governing various major agronomic traits of cultivated rice are known, but for identifying these QTLs, accurate phenotyping using multiple trials is required since the effects of QTLs are small and strongly affected by environmental variables (Datta 2005). Doubled haploids make this conceivable due to their true breeding nature and ease of mass production. Using DH as mapping population QTLs associated with resistance to brown planthopper in rice (Soundararajan et al. 2004), sheath blight resistance (Xu et al. 2011), root traits (Li et al. 2003), yield and its attributing traits (Hittalmani et al. 2002; Park et al. 2014), and protein content and lipid content for rice quality analysis (Lee et al. 2014) were identified which can be further used for enhancing the respective traits in rice.

5.3 Conclusions

The advancements in haploid rice and maize production methods demonstrate the revolutionary potential of haploid technology for crop improvement. Haploid production is positioned as a pillar for future developments in rice and maize breeding due to its capacity to produce homozygous lines quickly, introduce genetic variety, and use advanced genomic tools. In order to address global food security issues and promote sustainable agriculture, ongoing research efforts to improve and expand haploid production technologies will be essential.

References

Ahmadi B, Ebrahimzadeh H (2020) *In vitro* androgenesis: spontaneous vs. artificial genome doubling and characterization of regenerants. Plant Cell Rep 39:299–316

Ali J, Nicolas KLC, Akther S, Torabi A, Ebadi AA, Marfori-Nazarea CM, Mahender A (2021) Improved anther culture media for enhanced callus formation and plant regeneration in rice (Oryza sativa L.). Plan Theory 10:839

Asif M (2013) History, production methods and types of haploids. In: Asif M (ed) Progress and opportunities of double haploid production. University of Alberta, Edmonton, p 11

Baleroni AG, Re F, Pelozo A, Kamphorst SH, Carneiro JWP, Rossi RM, Scapim CA (2021) Identification of haploids and diploids in maize using seedling traits and flow cytometry. Crop Breed Appl Biotechnol 21(4):e38422145

Battistelli GM, Von Pinho RG, Justus A, Couto EG, Balestre M (2013) Production and identification of doubled haploids in tropical maize. Genet Mol Res 12:4230–4242

Bennett MD, Finch RA, Barclay IR (1976) The time rate and mechanism of chromosome elimination in Hordeum hybrids. Chromosoma 54(2):175–200

Bhat JG, Murthy HN (2007) Factors affecting in-vitro gynogenic haploid production in Niger (Guizotia abyssinica (L. f.) Cass.). Plant Growth Regul 52(3):241–248

Bishnoi U, Jain RK, Rohilla JS, Chowdhury VK, Gupta KR, Chowdhury JB (2000) Anther culture of recalcitrant Indica x basmati rice hybrids. Euphytica 114:93–101

Bjørnstad A, Opsahl-Ferstad HG, Aasmo M (1989) Effects of donor plant environment and light during incubation of anther cultures of some spring wheat (Triticum aestivum L.) cultivars. Plant Cell Tissue Organ Cult 17:27–37

Blakeslee AF (1922) Variations in Datura due to changes in chromosome number. Am Nat 56(642):16–31

Boote BW, Freppon DJ, De La Fuente GN, Lübberstedt T, Nikolau BJ, Smith EA (2016) Haploid differentiation in maize kernels based on fluorescence imaging. Plant Breed 135:439–445

Britt AB, Kuppu S (2016) CENH3: an emerging player in haploid induction technology. Front Plant Sci 7:357

Chaikam V (2012) In vivo maternal haploid induction in maize. In: Prasanna BM, Chaikam V, Mahuku G (eds) Doubled haploid technology in maize breeding: theory and practice. CIMMYT, Mexico, D.F

Chaikam V, Nair SK, Babu R, Martinez L, Tejomurtula J, Boddupalli PM (2015) Analysis of effectiveness of R1-nj anthocyanin marker for in vivo haploid identification in maize and molecular markers for predicting the inhibition of R1-nj expression. Theor Appl Genet 128:159–171

Chaikam V, Martinez L, Melchinger AE, Schipprack W, Boddupalli PM (2016) Development and validation of red root marker-based haploid inducers in maize. Crop Sci 56:1678–1688

Chaikam V, Lopez LA, Martinez L, Burgueño J, Boddupalli PM (2017) Identification of in vivo induced maternal haploids in maize using seedling traits. Euphytica 213:177

Chaikam V, Nair SK, Martinez L, Lopez LA, Utz HF, Melchinger AE, Boddupalli PM (2018) Marker-assisted breeding of improved maternal haploid inducers in maize for the tropical/subtropical regions. Front Plant Sci 9:Article 1527

Chaikam V, Molenaar W, Melchinger AE, Boddupalli PM (2019) Double haploid technology for line development in maize: technical advances and prospects. Theor Appl Genet 2019(132):3227–3243

Chaikam V, Molenaar W, Melchinger AE, Prasanna BM (2019a) Doubled haploid technology for line development in maize: technical advances and prospects. Theor Appl Genet 132:3227–3243

Chang M-T, Coe EH (2009) Doubled haploids. In: Molecular genetic approaches to maize improvement. Springer, pp 127–142

Chase SS (1947) Techniques for isolating monoploid maize plants. J Bot 34:582

Chase SS (1949) Spontaneous doubling of the chromosome complement in monoploid sporophytes of maize. Proc Iowa Acad Sci 56:113–115

Chase SS (1951) Production of homozygous diploids of maize from monoploids. Agron J 44(263):267

Chase SS (1952) Production of homozygous diploids of maize from monoploids. Agron J 44:263–267

Cha-um S, Srianan B, Pichakum A, Kirdmanee C (2009) An efficient procedure for embryogenic callus induction and double haploid plant regeneration through anther culture of Thai aromatic rice (Oryza sativa L. subsp. indica). In-Vitro Cell Dev Biol 45(2):171–179

Chen CC, Lin CM (1981) Genotypic differences in plant production in anther culture of rice. In: Chang WC (ed) Proceedings of symposium on plant tissue culture. Acad Sinica, Taipei, pp 199–203

Chen CC, Tsay HS, Huang CR (1991) Factors affecting androgenesis in rice (Oryza sativa L.). In: Bajaj YPS (ed) Biotechnology in agriculture and forestry. Springer, Berlin/Heidelberg, pp 193–215

Chen JF, Cui L, Malik AA, Mbira KG (2011) *In vitro* haploid and dihaploid production via unfertilized ovule culture. Plant Cell Tissue Organ Cult (PCTOC) 104(3):311–319

Chidzanga C, Muzawazi F, Midzi F, Hove T (2017) Production and use of haploids and doubled haploid in maize breeding: a review. Afr J Plant Breed 4(4):201–213

Choe E, Carbonero CH, Mulvaney K, Rayburn AL, Mumm RH (2012) Improving in vivo maize doubled haploid production efficiency through early detection of false positives. Plant Breed 131(3):399–401

Choudhary M, Wani SH, Kumar P, Bagaria PK, Rakshit S, Roorkiwal M, Varshney RK (2019) QTLian breeding for climate resilience in cereals: progress and prospects. Funct Integr Genomics 19:685–701

Chu CC (1978) The N6 medium and its applications to anther culture of cereal crops. In: Proceeding of symposium on plant tissue culture, pp 43–50

Cistue L, Ramos A, Castilo AM, Romagosa I (1994) Production of large number of doubled haploid plants from barley anthers pretreated with high concentrations of mannitol. Plant Cell Rep 13(12):709–712

Coe EH (1959) A line of maize with high haploid frequency. Am Nat 93:381–382

Couto EG, Davide LMC, Bustamante FO, Pinho RGV, Silva TN (2013) Identification of haploid maize by flow cytometry, morphological and molecular markers. Cienc Agrotec 37:25–31

Cowen NM, Johnson CD, Armstrong K, Miller M, Woosley A, Pescitelli S, Skokut M, Belmar S, Petolino JF (1992) Mapping genes conditioning *in vitro* androgenesis in maize using RFLP analysis. Theor Appl Genet 84:720–724

Cui YC, Schmitzer PR, Young DH (2013) Induced chromosome doubling in plants. Google Patents

Dang NC, Munsch M, Aulinger I, Renlai W, Stamp P (2012) Inducer line generated double haploid seeds for combined waxy and opaque 2 grain quality in subtropical maize (Zea mays. L.). Euphytica 183:153–160

Datta SK (2001) Androgenesis in cereals. In: Bhojwani SS, Soh WY (eds) Current trends in the embryology of angiosperms. Springer, Dordrecht, pp 471–488

Datta SK (2005) Androgenic haploids: factors controlling development and its application in crop improvement. Curr Sci 89:1870–1878

Datta SK, Wenzel G (1998) Single microspore derived embryogenesis and plant formation in barley (Hordeum vulgare). Arch Zeucht 18:125–131

Davies DR (1974) Chromosome elimination in inter-specific hybrids. Heredity 32(2):267–270

Davoyan ZI (1985) Production of rice haploids by cultivating infertile ovaries *in vitro*. Sov Agric Sci 3:21–23

De La Fuente GN, Carstensen JM, Edberg MA, Lübberstedt T (2017) Discrimination of haploid and diploid maize kernels via multispectral imaging. Plant Breed 136:50–60

De Souza IRP, Dos Trindade SR, De Vasconcelos MJV, De Paula ALSP, De Guimarães OPE, Guimarães LJM, De Alves CM (2018) Estratégias para identificação de falso-positivos no processo de obtenção de duplo-haploides em milho. Embrapa Milho e Sorgo, Sete Lagoas, p 20. (Embrapa Milho e Sorgo. Boletim de Pesquisa e Desenvolvimento, 171)

Devaux P, Pickering R (2005) Haploids in the improvement of Poaceae. In: Palmer CE, Keller WA, Kasha KJ (eds) Haploids in crop improvement II. Biotechnology in agriculture and forestry, vol 56. Springer, Berlin/New York, pp 215–242

Dong L, Li L, Liu C, Liu C, Geng S, Li X, Xie C (2018) Genome editing and double-fluorescence proteins enable robust maternal haploid induction and identification in maize. Mol Plant 11:1214

Dunwell JM (2010) Haploids in flowering plants: origins and exploitation. Plant Biotechnol J 8(4):377–424

Dwivedi SL, Britt AB, Tripathi L, Sharma S, Upadhyaya HD, Ortiz R (2015) Haploids: constraints and opportunities in plant breeding. Biotechnol Adv 33(6):812–829

Faruq G, Shamsuddin F, Jenifer A, Jusoh SH, Nezhadahmadi A, Khalid N (2014) Optimization of media and cold pre-treatment for anther culture using japonica/indica and indica/indica hybrids and their callus induction comparisons in different rice crosses. Ind J Sci Technol 7(11):1861–1870

Finch RA (1983) Tissue-specific elimination of alternative whole parental genomes in one barley hybrid. Chromosoma 88(5):386–393

Foiada F, Westermeier P, Kessel B, Ouzunova M, Wimmer V, Mayerhofer W, Presterl T, Dilger M, Kreps R, Eder J, Schön C-C (2015) Improving resistance to the European corn borer: a comprehensive study in elite maize using QTL mapping and genome-wide prediction. Theor Appl Genet 128:875–891

Foroughi-Wehr B, Wenzel G (1993) Andro- and parthenogenesis. In: Hayward MD, Bosemark NO, Romagosa I, Cerezo M (eds) Plant breeding. Springer, Dordrecht, pp 261–277

Forster BP, Thomas WTB (2005) Doubled haploids in genetics and plant breeding. Plant Breed Rev 25:57–88

Forster BP, Thomas WTB (2010) Doubled haploids in genetics and plant breeding. In: Janick J (ed) Plant breeding reviews. Wiley, Oxford, pp 57–88

Forster BP, Heberle-Bors E, Kasha KJ, Touraev A (2007) The resurgence of haploids in higher plants. Trends Plant Sci 12(8):368–375

Garrido D, Vicente O, Heberle-Bors E, Rodriguez-Garcia MI (1995) Cellular changes during the acquisition of embryogenic potential in isolated pollen grains of Nicotiana tabacum. Protoplasma 186:220–230

Gayen P, Madan JK, Kumar R, Sarkar KR (1994) Chromosome doubling in haploids through colchicine. Maize Genet Coop Newsl 68:65

Geiger HH (2009) Doubled haploids. In: Bennetzen JL, Hake S (eds) Maize handbook. Vol. II: Genetics and genomics, vol v. 2. Springer, Heidelberg/New York, pp 641–659

Geiger HH, Deimling S (1994) Herbicide resistance as a marker in screening for maternal haploids. Maize Genet Coop News Lett 68

Genovesi A, Collins G (1982) *In vitro* production of haploid plants of corn via anther culture. Crop Sci 22:1137–1144

Gosal SS, Sindhu AS, Sandhu JS, Sandhu-Gill R, Singh B, Khehra GS, Dhaliwal HS (1997) Haploidy in rice. In: Jain SM, Sopory SK, Veilleux RE (eds) *In vitro* haploid production in higher plants. Springer, Dordrecht, pp 1–35

Grewal D (2009) Doubled haploids: from laboratory to field. Rice Today 8(3):46

Grewal D, Manito C, Bartolome V (2011) Doubled haploids generated through anther culture from crosses of elite Indica and Japonica cultivars and/or lines of rice: large-scale production, agronomic performance, and molecular characterization. Crop Sci 51(6):2544–2553

Gueye T, Ndir KN (2010) *In vitro* production of double haploid plants from two rice species (Oryza sativa L. and Oryza glaberrima Steudt.) for the rapid development of new breeding material. Sci Res Essays 5:709–713

Guha S, Maheshwari SC (1964) *In vitro* production of embryos from anthers of Datura. Nature 204(4957):497–497

Guha-Mukherjee S (1973) Genotypic differences in the *in vitro* formation of embryoids from rice pollen. J Exp Bot 24:139–144

Gupta M, Choudhary M, Kumar H, Kaswan V, Kaur Y, Choudhary JR, Yadav S (2021) Doubled haploid technology in maize (*Zea mays*): status and applications. Indian J Agric Sci 92(3):00–00

Gustin JL, Frei UK, Baier J, Armstrong P, Lübberstedt T, Settles AM (2020) Classification approaches for sorting maize (Zea mays subsp. mays) haploids using single-kernel near-infrared spectroscopy. Plant Breed 139(6):1103–1112

He T, Yang Y, Tu SB, Yu MQ, Li XF (2006) Selection of interspecific hybrids for anther culture of Indica rice. Plant Cell Tissue Organ Cult 86:271–277

Heberle-Bors E (1985) *In vitro* haploid formation from pollen: a critical review. Theor Appl Genet 71:361–374

Heberle-Bors E (1989) Isolated pollen culture in tobacco: plant reproductive development in a nutshell. Sex Plant Reprod 2:1–10

Herath HMI, Bandara DC (2011) Anther culture performance in selected high yielding Indica (of Sri Lanka) and Japonica rice varieties and their inter sub-specific hybrids. J Natl Sci Found 39:149–154

Herath HMI, Bandara DC, Samarajeewa PK (2007) Effect of culture media for anther culture of indica rice varieties and hybrids of Indica and Japonica. Trop Agric Res Ext 10:17–22

Herath HMI, Bandara DC, Samarajeewa PK, Wijesundara DSA (2009) Effect of low temperature pre-treatment on anther culture in selected indica, japonica rice varieties and their inter subspecific hybrids. Cey J Sci 38:11–16

Hirsch CD, Wu Y, Yan H, Jiang J (2009) Lineage-specific adaptive evolution of the centromeric protein CENH3 in diploid and allotetraploid Oryza species. Mol Biol Evol 26(12):2877–2885

Hittalmani S, Shashidhar HE, Bagali PG, Huang H, Sidhu JS, Singh VP, Khush GS (2002) Molecular mapping of quantitative trait loci for plant growth, yield and yield related traits across three diverse locations in a doubled haploid rice population. Euphytica 125(2):207–214

Hu H, Schrag TA, Peis R, Unterseer S, Schipprack W, Chen S (2016) The genetic basis of haploid induction in maize identified with a novel genome-wide association method. Genetics 202:1267–1276

Ikeuchi M, Favero DS, Sakamoto Y, Iwase A, Coleman D, Rymen B, Sugimoto K (2019) Molecular mechanisms of plant regeneration. Annu Rev Plant Biol 70:377–406. https://doi.org/10.1146/annurev-arplant-050718-100434

Ingle PB, Rokade SS, Raut RV (2023) Review on double haploid in rice plant (Oryza sativa L.). J Alumni Assoc Biotechnol 5(1):1–4

Jacquard C, Asakaviciute R, Hamalian AM, Sangwan RS, Devaux P, Clement C (2006) Barley anther culture: effects of annual cycle and spike position on microspore embryogenesis and albinism. Plant Cell Rep 25:375–381

Jacquier NM, Gilles LM, Pyott DE, Martinant JP, Rogowsky PM, Widiez T (2020) Puzzling out plant reproduction by haploid induction for innovations in plant breeding. Nat Plants 6(6):610–619

Jang JH, Noh G, Seo HS, Jung KH, Kim YJ, Lee OR (2023) Loss of function of pollen-expressed phospholipase OsMATL2 triggers haploid induction in japonica rice. Plant Physiol 193(3):1749–1752

Javed MA, Samad AA, Hidayat T, Wagiran A, Baqr ARW (2012) Studies to investigate the interactions of genotypes, culture media and culture temperatures on androgenesis in recalcitrant indica rice (Oryza sativa L.). J Teknol 59(1):75–79

Jones RW, Reinot T, Frei UK, Tseng Y, Lubberstedt T, Mcclelland JF (2012) Selection of haploid maize kernels from hybrid kernels for plant breeding using near-infrared spectroscopy and SIMCA analysis. Appl Spectrosc 66(4):447–450

Kalinowska K, Chamas S, Unkel K, Demidov D, Lermontova I, Dresselhaus T, Houben A (2019) State-of-the-art and novel developments of in vivo haploid technologies. Theor Appl Genet 132(3):593–605

Kao KN (1981) Plant formation from barley anther cultures with ficoll media. J Plant Physiol 103(5):437–443

Kasha KJ, Kao KN (1970) High frequency haploid production in barley (Hordeum vulgare L.). Nature 225(5235):874–876

Kasha KJ, Maluszynski M (2003) Production of doubled haploids in crop plants: an introduction. In: Maluszynski M, Kasha K, Forster BP, Szarejko I (eds) Doubled haploid production in crop plants: a manual. Springer, New York, pp 1–4

Kato A (2002) Chromosome doubling of haploid maize seedlings using nitrous oxide gas at the flower primordial stage. Plant Breed 121:370–377

Kato A, Geiger HH (2002) Chromosome doubling of haploid maize seedlings using nitrous oxide gas at the flower primordial stage. Plant Breed 121:370–377

Kaur K, Neelam K, Singh J, Malik P, Singh K (2023) Uncovering natural allelic and structural variants of OsCENH3 gene by targeted resequencing and in silico mining in genus Oryza. Sci Rep 13:830

Kaushal L, Sharma R, Balachandran SM, Ulaganathan K, Shenoy V (2014) Effect of cold pre-treatment on improving anther culture response of rice (Oryza sativa L.). J Exp Biol Agric Sci 2(2S):234–242

Keller EJ, Korzun L (1996) Ovary and ovule culture for haploid production. In: Jain SM, Sopory SK, Veilleux RE (eds) *In vitro* haploid production in higher plants. Springer, Dordrecht, pp 217–235

Kelliher T, Starr D, Richbourg L, Chintamanani S, Delzer B, Nuccio ML, Green J, Chen Z, McCuiston J, Wang W, Liebler T, Bullock P, Martin B (2017) MATRILINEAL, a spermspecific phospholipase, triggers maize haploid induction. Nature 542:105–109

Kelliher T, Starr D, Su X, Tang G, Chen Z, Carter J, Wittich PE, Dong S, Green J, Burch E, McCuiston J, Gu W, Sun Y, Strebe T, Roberts J, Bate NJ, Que Q (2019) One-step genome editing of elite crop germplasm during haploid induction. Nat Biotechnol 37:287–292

Kermicle JL (1969) Androgenesis conditioned by a mutation in maize. Science 166:1422–1424

Kermicle JL (1994) Indeterminate gametophyte (ig): biology and use. In: Freeling M, Walbot V (eds) The maize handbook. Springer, New York, pp 388–393

Khulbe RK, Pattanayak A (2021) Genetics of *in vivo* maternal haploid induction in maize. In: Doubled haploid production in maize using *in vivo* maternal haploid induction system. ICAR-Vivekananda Parvatiya Krishi Anusandhan Sansthan, Almora. v + 60

Kiviharju E, Pehu E (1998) The effect cold and heat pretreatments on anther culture response of Avena sativa and A. sterilis. Plant Cell Tissue Organ Cult 54:97–104

Kleiber D, Prigge V, Melchinger AE et al (2012) Haploid fertility in temperate and tropical maize germplasm. Crop Sci 52:623–630

Kumar P, Choudhary M, Hossain F, Singh N, Choudhary P, Gupta M, Singh V, Chikappa G, Kumar R, Kumar B, Jat S, Rakshit S (2019) Nutritional quality improvement in maize (Zea mays): progress and challenges. Indian J Agric Sci 89(6):8950–8911

Kumar K, Gambhir G, Dass A, Tripathi AK, Singh A, Jha AK, Yadava P, Choudhary M, Rakshit S (2020) Genetically modified crops: current status and future prospects. Planta 251:1–27

Kumari S (2019) Production of doubled haploids in rice (Oryza sativa L.) through anther culture and wide crosses and their molecular characterization. Doctoral dissertation, Punjab Agricultural University, Ludhiana

Kumlehn J (2014) Haploid technology. In: Kumlehn J, Stein N (eds) Biotechnological approaches to barley improvement. Biotechnology in agriculture and forestry, vol 69. Springer, Heidelberg, pp 379–392

Kuo CS (1982) The preliminary studies on culture of unfertilized ovary of rice *in vitro*. Acta Bot Sin 24:33–38

Kyum M, Kaur H, Kamboj A, Goyal L, Bhatia D (2021) Strategies and prospects of haploid induction in rice (Oryza sativa), vol 141. Plant Breed, pp 1–11

Lapitan PC, Violeta NV (1999) Anther culture response of some genotypes with varying amylase content to nitrogen fixation. Phill J Crop Sci 24(3):79–84

Lashermes P (1992) Improved anther culture method for obtaining direct regeneration in wheat (Triticum aestivum L.). J Genet Breed 46:99–102

Lee JH, Lee SY (1995) Effect of gelling agents and growth regulators on rice anther culture. Kor J Plant Tiss Cult 22(1):35–39

Lee JH, Lee SY (2002) Selection of stable mutants from cultured rice anthers treated with ethyl methane sulfonic acid. Plant Cell Tissue Organ Cult 71(2):165–171

Lee GH, Yun BW, Kim KM (2014) Analysis of QTLs associated with the rice quality related gene by double haploid populations. Int J Genom 2014:1–6

Lentini Z, Reyes P, Martinez CP, Roca WM (1995) Androgenesis in highly recalcitrant rice genotypes with maltose and silver nitrate. Plant Sci 110:127–138

Li ZK, Yu SB, Lafitte HR, Huang N, Courtois B, Hittalmani S et al (2003) QTL × environment interactions in rice: I. Heading date and plant height. Theor Appl Genet 108(1):141–153

Li L, Xu X, Jin W, Chen S (2009) Morphological and molecular evidences for DNA introgression in haploid induction via a high oil inducer CAUHOI in maize. Planta 230(2):367–376

Liu C, Li W, Zhong Y, Dong X, Hu H, Tian X, Wang L, Chen B, Chen C, Melchinger AE, Chen S (2015) Fine mapping of qhir8 affecting in vivo haploid induction in maize. Theor Appl Genet 128:2507–2515

Liu Z, Wang Y, Ren J, Mei M, Frei UK, Trampe B, Lübberstedt T (2016) Maize doubled haploids. In: Janick J (ed) Plant breeding reviews, vol 40, 1st edn. Wiley-Blackwell

Liu C, Li X, Meng D, Zhong Y, Chen C, Dong X, Tian X (2017) A 4-bp insertion at ZmPLA1 encoding a putative phospholipase A generates haploid induction in maize. Mol Plant 10:520–522

Liu J, Liang D, Yao L, Zhang Y, Liu C, Liu Y, Wang Y, Zhou H, Kelliher T, Zhang X et al (2021) Rice haploid inducer development by genome editing. In: Bandyopadhyay A, Thilmony R

(eds) Rice genome engineering and gene editing: methods in molecular biology, vol 2238. Humana, New York, pp 221–230
Lv J, Yu K, Wei J, Gui H, Liu C, Liang D, Kelliher T (2020) Generation of paternal haploids in wheat by genome editing of the centromeric histone CENH3. Nat Biotechnol 38(12):1397–1401
Ma H, Li G, Würschum T, Zhang Y, Zheng D, Yang X, Li J, Liu W, Yan J, Chen S (2018) Genome-wide association study of haploid male fertility in maize (Zea mays L.). Front Plant Sci 9:Article 974
Mandal N, Gupta S (1995) Effect of genotype and culture medium on androgenic callus formation and green plant regeneration in indica rice. Ind J Exp Biol 33:761–765
Marrone BL (2009) Flow cytometry: a multipurpose technology for a wide spectrum of global biosecurity applications. J Lab Autom 14:148–156
Martínez R, Cesar P, Arumuganathan K, Kikuchi H, Earle E (1994) Nuclear DNA content of ten rice species as determined by flow cytometry. Jpn J Genet 69:513–523
Mayakaduwa R, Silva T (2023) Haploid induction in indica rice: exploring new opportunities. Plan Theory 12:3118
Medhabati K, Das KR, Henary C, Singh TD, Sunitibala H (2014) Androgenic callus induction of the indica rice hybrid of Chakhao Amubi and Basmati 370. Int Res J Biol Sci 3(4):73–79
Melchinger AE, Schipprack W, Würschum T, Chen S, Technow F (2013) Rapid and accurate identification of in vivo-induced haploid seeds based on oil content in maize. Sci Rep 3:2129
Melchinger AE, Schipprack W, Friedrich Utz H, Mirdita V (2014) In vivo haploid induction in maize: identification of haploid seeds by their oil content. Crop Sci 54:1497–1504
Melchinger AE, Schipprack W, Mi X, Mirdita V (2015) Oil content is superior to oil mass for identification of haploid seeds in maize produced with high-oil inducers. Crop Sci 55:188–195
Melchinger AE, Molenaar WS, Mirdita V, Schipprack W (2016) Colchicine alternatives for chromosome doubling in maize haploids for doubled-haploid production. Crop Sci 56:559–569
Melchinger AE, Brauner PC, Böhm J, Schipprack W (2016b) In vivo haploid induction in maize: comparison of different testing regimes for measuring haploid induction rates. Crop Sci 56:1127–1135
Mirdita V, Schipprack W, Melchinger A (2014) Identification of in vivo induced haploid seeds in maize based on oil content. 50th Annual Illinois Corn Breeders School, Illinois
Mishra R, Rao GJN (2016) In-vitro androgenesis in rice: advantages, constraints and future prospects. Rice Sci 23(2):57–68
Mishra R, Rao GJN, Rao RN (2013) Effect of cold pretreatment and phytohormones on anther culture efficiency of two indica rice (Oryza sativa L.) hybrids: Ajay and Rajalaxmi. J Exp Biol Agric Sci 1(2):69–76
Mishra R, Rao GJN, Rao RN, Kaushal P (2015) Development and characterization of elite-doubled haploid lines from two indica rice hybrids. Rice Sci 22(6):290–299
Mishra R, Jwala G, Rao N (2016) *In-vitro* androgenesis in rice: advantages, constraints and future prospects. Rice Sci 23(2):57–68
Mohiuddin AKM, Karim NH, Shahanaz S (2014) Development of improved doubled-haploids through anther culture of indica rice (Oryza sativa L.). Ann Biol Res 5:6–13
Molenaar WS, Schipprack W, Melchinger AE (2018) Nitrous oxideinduced chromosome doubling of maize haploids. Crop Sci 58:650–659
Molenaar WS, Schipprack W, Brauner PC, Melchinger AE (2019) Haploid male fertility and spontaneous chromosome doubling evaluated in a diallel and recurrent selection experiment in maize. Theor Appl Genet 132:2273
Mukherjee A, Islam MR, Nasiruddin KM, Banerjee P (2015) Study on callus initiation and plantlet regeneration ability of some rice genotypes. Int J Sci Technol Res 4(10):354–361
Murashige T, Skoog F (1962) A revised medium for rapid growth and bioassays with tobacco tissue cultures. Physiol Plant 15:473–497
Murovec J, Bohanec B (2012) Haploids and doubled haploids in plant breeding. In: Abdurakhmonov I (ed) Plant breeding. IntechOpen, Rijeka, pp 87–106
Myint SK, Aung KO, Soe K (2005) Development of a short duration upland rice mutant line through anther culture of gamma irradiated plants. Mutat Breed Newsl Rev 1:13–14

Nagaki K, Cheng Z, Ouyang S, Talbert PB, Kim M, Jones KM, Jiang J (2004) Sequencing of a rice centromere uncovers active genes. Nat Genet 36(2):138–145

Naik N, Rout P, Umakanta N, Verma RL, Katara JL, Sahoo KK et al (2017) Development of doubled haploids from an elite indica rice hybrid (BS6444G) using anther culture. Plant Cell Tissue Organ Cult 128(3):679–689

Nanda DK, Chase SS (1966) An embryo marker for detecting monoploids of maize (Zea mays L.). Crop Sci 6:213–215

Nathanaël MA, Jacquier Laurine M, Gilles Jean-Pierre, Martinant Peter M, Rogowsky Thomas, Widiez Jose M, Segui-Simarro (2021) Doubled Haploid Technology, Volume 2: Hot Topics Apiaceae Brassicaceae Solanaceae. Maize In Planta Haploid Inducer Lines: A Cornerstone for Doubled Haploid Technology. Springer US New York NY 25–48

Niizeki H, Oono K (1968) Induction of haploid rice plant from anther culture. Proc Jpn Acad 44(6):554–557

Niroula RK, Bimb HP (2009) Effect of genotype and callus induction medium on green plant regeneration from anther of Nepalese rice cultivars. Asian J Plant Sci 8(5):368–374

Nurhasanah, Pratama AN, Sunaryo W (2016) Anther culture of local upland rice varieties from East Kalimantan: effect of panicle cold pre-treatment and putrescine-enriched medium. Biodiversitas 17(1):148–153

Ogawa T, Fukuwa H, Ohakawa Y (1995) Plant regeneration through direct culture of isolated pollen grains in rice. Breed Sci 45(3):301–307

Pande H (1997) Androgenesis in anther cultures of an indica cultivar of Oryza sativa L. Ph.D. thesis, University of Delhi, Delhi, India

Park SG, Ubaidillah M, Kim KM (2013) Effect of maltose concentration on plant regeneration of anther culture with different genotypes in rice (Oryza sativa L.). Am J Plant Sci 4(11):2265–2270

Park GH, Kim JH, Kim KM (2014) QTL analysis of yield components in rice using a cheong-cheong/nagdong doubled haploid genetic map. Am J Plant Sci 5(9):1174–1180

Patra S, Mohanty P, Tripathy SK (2019) Scope of androgenesis for genetic improvement in rice. Emerg Life Sci Res 5(1):23–28

Pollacsek M (1992) Management of the ig gene for haploid induction in maize. Agronomie 12:247–251

Prasanna B, Chaikam V, Mahuku G (2012) Doubled haploid technology in maize breeding: theory and practice. CIMMYT, Mexico, DF, p 50

Prem D, Gupta K, Agnihotri A (2004) Doubled haploids: a powerful biotechnological tool for genetic enhancement in oilseed brassicas. In: Srivastava PS, Narula A, Srivastava S (eds) Plant biotechnology and molecular markers. Anamaya Publishers, Delhi, pp 18–30

Prigge V, Melchinger AE (2011) Production of haploids and doubled haploids In maize. In: Loyola-Vargas VM, Ochoa-Alejo N (eds) Plant cell culture protocols, 3rd edn. Humana, Totowa

Prigge V, Xu X, Li L, Babu R, Chen S, Atlin GN, Melchinger AE (2012) New insights into the genetics of in vivo induction of maternal haploids, the backbone of doubled haploid technology in maize. Genetics 190:781–793

Qian Q, Zeng D, He P, Zheng X, Chen Y, Zhu L (2000) QTL analysis of the rice seedling cold tolerance in a double haploid population derived from anther culture of a hybrid between indica and japonica rice. Chin Sci Bull 45(5):448–453

Qiu F, Liang Y, Li Y et al (2014) Morphological, cellular and molecular evidences of chromosome random elimination in vivo upon haploid induction in maize. Curr Plant Biol 1:83–90

Raina SK (1989) Tissue culture in rice improvement: status and potential. Adv Agron 42:339–398

Raina SK, Irfan ST (1998) High-frequency embryogenesis and plantlet regeneration from isolated microspores of indica rice. Plant Cell Rep 17(12):957–962

Randolph LF (1932) Some effects of high temperature on polyploidy and other variations in maize. Genetics 18:222–29

Rao PS, Suprasanna P (1996) Methods to double haploid chromosome numbers. In: Jain SM, Sopory SK, Veilleux RE (eds) In vitro haploid production in higher plants. Current Plant Science and Biotechnology in Agriculture, vol 23. Springer, Dordrecht. pp 317–339. https://doi.org/10.1007/978-94-017-1860-8_18

Ratheika V, Silva TD (2007) A study on the anther culture response in different varieties of rice (Oryza sativa L.) subspecies indica. In: Proceedings of the 27th annual sessions of the Institute of Biology, Sri Lanka, p 28

Ravi M, Chan SW (2010) Haploid plants produced by centromere mediated genome elimination. Nature 464(7288):615–618

Reinert J, Bajaj YPS (1977) Anther culture: haploid production and its significance. In: Reinert J, Bajaj YPS (eds) Applied and fundamental aspects of plant cell, tissue and organ culture. Springer, Berlin/Heidelberg, pp 251–267

Ren J, Wu P, Tian X et al (2017) QTL mapping for haploid male fertility by a segregation distortion method and fine mapping of a key QTL qhmf4 in maize. Theor Appl Genet 130:1349–1359

Ren J, Boerman NA, Liu R, Wu P, Trampe B, Vanous K, Frei UK, Chen S, Lübberstedt T (2020) Mapping of QTL and identification of candidate genes conferring spontaneous haploid genome doubling in maize (Zea mays L.). Plant Sci 293:110337

Rober F, Gordillo G, Geiger H (2005) In vivo haploid induction in maize-performance of new inducers and significance of doubled haploid lines in hybrid breeding. Maydica 50:275

Rotarenco V, Kirtoca I, Jacota A (2007) Using oil content to identify kernels with haploid embryos. Maize Genet Coop News Lett 81:11

Rotarenco V, Dicu G, Fuia S (2010) New inducers of maternal haploids in maize. Maize Genet Coop News Lett:21–22

Rout P, Naik N, Ngangkham U, Verma RL, Katara JL, Singh ON, Samantaray S (2016) Doubled haploids generated through anther culture from an elite long-duration rice hybrid, CRHR32: method optimization and molecular characterization. Plant Biotechnol 33(3):177–186

Sah BP (2008) Response of genotypes to culture media for callus induction and regeneration of plants from rice anthers. Sci World 6(6):37–43

Samantaray S, Ali J, Nicolas KLC, Katara JL, Verma RL, Parameswaran C, Devanna BN, Kumar A, Dash B, Bhuyan SS (2021) Doubled haploids in rice improvement: approaches, applications, and future prospects. In: Ali J, Wani SH (eds) Rice improvement: physiological, molecular breeding and genetic perspectives. Springer, Cham, pp 425–447

Sanei M, Pickering R, Kumke K, Nasuda S, Houben A (2011) Loss of centromeric histone H3 (CENH3) from centromeres precedes uniparental chromosome elimination in interspecific barley hybrids. Proc Natl Acad Sci 108(33):E498–E505

Santel H (2009) Laudis® OD–a new herbicide for selective post-emergence weed control in corn (Zea mays L.). Bayer Crop Sci J 62:95–108

Saurabh S (2021) Genome editing: revolutionizing the crop improvement. Plant Mol Biol Report 39:1–21

Sen C, Singh RP, Singh MK, Singh HB (2011) Effect of cold pretreatment on anther culture of boro rice hybrids. Int J Plant Rep Biol 3:69–73

Senadhira D, Zapata-Arias F, Gregorio G, Alejar M, de la Cruz H, Padolina T, Galvez A (2002) Development of the first salt-tolerant rice cultivar through indica/indica anther culture. Field Crop Res 76:103–110

Shariatpanahi ME, Bal U, Heberle-Bors E, Touraev A (2006) Stresses applied for the re-programming of plant microspores towards *in vitro* embryogenesis. Physiol Plant 127:519–534

Silva TD (2010) Indica rice anther culture: can the impasse be surpassed? Plant Cell Tissue Organ Cult 100(1):1–11

Silva TD (2012) Microspore embryogenesis. In: Sato KI (ed) Embryogenesis. IntechOpen, Rijeka, pp 573–596

Silva TD, Ratnayake WJ (2009) Anther culture potential of indica rice varieties, Kurulu Thuda and BG250. Trop Agric Res Ext 12(2):53–56

Smykal P (2000) Pollen embryogenesis – the stress mediated switch from gametophytic to sporophytic development: current status and future prospects. Biol Plant 43:481–489

Song S, Tian D, Zhang Z, Hu S, Yu J (2019) Rice genomics: over the past two decades and into the future. Genom Proteomics Bioinf 1:1

Soundararajan RP, Kadirvel P, Gunathilagaraj K, Maheswaran M (2004) Mapping of quantitative trait loci associated with resistance to brown planthopper in rice by means of a doubled haploid population. Crop Sci 44:2214–2220

Strigens A, Schipprack W, Reif JC, Melchinger AE (2013) Unlocking the genetic diversity of maize landraces with doubled haploids opens new avenues for breeding. PLoS One 8(2):e57234

Szarejko I (2003) Anther culture for double haploid production in barley (Hordeum valgare L.). In: Maluszynski M, Kasha K, Forster BP, Szarejko I (eds) Double haploid production in crop plants: a manual. Kluwer Academic Publishers, Dordrecht, pp 35–42

Talebi R, Rahemi MR, Arefi H, Nourozi M, Bagheri N (2007) *In vitro* plant regeneration through anther culture of some Iranian local rice (Oryza sativa L.) cultivars. Pak J Biol Sci 10(12):2056–2060

Tang F, Tao Y, Zhao T, Wang G (2006) *In vitro* production of haploid and doubled haploid plants from pollinated ovaries of maize (Zea mays). Plant Cell Tissue Organ Cult 84:233–237

Testillano P, Georgiev S, Mogensen HL, Coronado MJ, Dumas C, Risueno MC, Matthys-Rochon E (2004) Spontaneous chromosome doubling results from nuclear fusion during *in vitro* maize induced microspore embryogenesis. Chromosoma 112:342–349

Tian X, Qin Y, Chen B, Liu C, Wang L, Li X, Dong X, Liu L, Chen S (2018) Hetero-fertilization together with failed egg–sperm cell fusion supports single fertilization involved in in vivo haploid induction in maize. J Exp Bot 69(20):4689–4701

Touraev A, Vicente O, Heberle-Bors E (1997) Initiation of microspore embryogenesis by stress. Trends Plant Sci 2:297–302

Tran DG, Vuong DT (2002) Effect of different media and genotypes on anther culture efficiency of F1 plants derived from crosses between IR64 and new plant type rice cultivars. Omonrice 10:107–109

Trejo-Tapia G, Amaya UM, Morales GS, Sanchez ADJ, Bonfil BM, Rodriguez-Monroy M, Jimenez-Aparicio A (2002) The effects of cold-pretreatment, auxins and carbon source on anther culture of rice. Plant Cell Tissue Organ Cult 71:41–46

Trindade RS (2022) The use of doubled haploid technology for lines development in maize breeding. Braz J Maize Sorghum 21:e1289

Tripathy SK, Swain D, Mohapatra PM, Prusti AM, Sahoo B, Panda S, Dash M, Chakma B, Behera SK (2019a) Exploring factors affecting anther culture in rice (*Oryza sativa* L.). J Appl Biol Biotechnol 7(02):87–92

Tripathy SK, Lenka D, Prusti AM, Mishra D, Swain D, Behera SK (2019b) Anther culture in rice: progress and breeding perspective. Appl Biol Res 21(2):87

Tseng Y (2012) New methods for haploid selection in maize. Thesis. Iowa State University of Science and Technology, Iowa, USA. https://doi.org/10.31274/etd-180810-256

Tsonev S, Todorovska E, Avramova V, Kolev S, Abu-Mhadi N, Christov NK (2009) Genomics assisted improvement of drought tolerance in maize: QTL approaches. Biotechnol Biotechnol Equip 23:1410–1413

Tuvesson S, Dayteg C, Hagberg P, Manninen O, Tanhuanpa P, Tenhola-Roininen T, Kiviharju E, Weyen J, Forster J, Schondelmaier J, Lafferty J, Marn M, Fleck A (2007) Molecular markers and doubled haploids in European plant breeding programmes. Euphytica 158(3):305–312

Veeraraghavan R (2007) A study on the comparison of anther culture response in different varieties of rice (Oryza sativa L.) subspecies indica. University of Colombo, Colombo. PMCid: PMC1891355

Wan Y, Duncan DR, Rayburn AL, Petolino JF, Widholm JM (1991) The use of antimicrotubule herbicides for the production of doubled haploid plants from anther-derived maize callus. Theor Appl Genet 81:205–211

Wang CC, Sun CS, Chu CC, Wu SC (1978) Studies on the albino pollen plantlets of rice. In: Proceeding of symposium on anther culture. Science Press, Beijing, pp 141–150

Wang Z, Taramino G, Yong D, Liu G, Tingey SV, Miao GL, Wang GL (2001) Rice ESTs with disease-resistance gene- or defense-response gene-like sequences mapped to regions containing major resistance genes or QTLs. Mol Genet Genomics 265(2):302–310

Wang B, Zhu L, Zhao B, Zhao Y, Xie Y, Zheng Z, Wang H (2019a) Development of a haploid-inducer mediated genome editing system for accelerating maize breeding. Mol Plant 12:597–602

Wang C, Liu Q, Shen Y, Hua Y, Wang J, Lin J, Wu M, Sun T, Cheng Z, Mercier R et al (2019b) Clonal seeds from hybrid rice by simultaneous genome engineering of meiosis and fertilization genes. Nat Biotechnol 37:283–286

Wang J, Cao Y, Wang K, Liu C (2022) Development of multiple-heading-date mtl haploid inducer lines in rice. Agriculture 12:806

Wang J, Yan H, Jiao X, Ren J, Hu F, Liang H, Liang W, Liu C (2023) Development of specific molecular and phenotypic marker-based haploid inducers in rice. Agronomy 13:1520

Wani SH, Choudhary M, Kumar P, Akram NA, Surekha C, Ahmad P, Gosal SS (2018) Marker-assisted breeding for abiotic stress tolerance in crop plants. In: Biotechnologies of crop improvement, vol 3. Springer, Cham, pp 1–23

Watts A, Kumar V, Bhat SR (2016) Centromeric histone H3 protein: from basic study to plant breeding applications. J Plant Biochem Biotechnol 25(4):339–348

Watts A, Kumar V, Raipuria RK, Bhattacharya RC (2018) In vivo haploid production in crop plants: methods and challenges. Plant Mol Biol Report 36(5):685–694

Weber DF (2014) Chapter three-today's use of haploids in corn plant breeding. Adv Agron 123:123–144

Wetherell DF (1984) Enhanced adventitive embryogenesis resulting from plasmolysis of cultured wild carrot cells. Plant Cell Tissue Organ Cult 3:221–227

William KJ, Taylor SP, Bogacki P, Pallotta M, Bariana HS, Wallwork H (2002) Mapping of the root lesion nematode (Pratylenchus neglectus) resistance gene Rlnn 1 in wheat. Theor Appl Genet 104:874–879

Wu P, Ren J, Tian X et al (2017) New insights into the genetics of haploid male fertility in maize. Crop Sci 57:637–647

Xa TT, Lang NT (2011) Rice breeding for high grain quality through anther culture. Omonrice 18:68–72

Xu Q, Yuan XP, Yu HY, Wang YP, Tang SX, Wei XH (2011) Mapping quantitative trait loci for sheath blight resistance in rice using double haploid population. Plant Breed 130(3):404–406

Yan J, Xue Q, Zhu J (1996) Genetic studies of anther culture ability in rice (Oryza sativa). Plant Cell Tissue Organ Cult 45:253–258

Yang HY, Zhou C (1982) *In vitro* induction of haploid plants from unpollinated ovaries and ovules. Theor Appl Genet 63(2):97–104

Yang Y, Zhu K, Xia H, Chen L, Chen K (2014) Comparative proteomic analysis of Indica and Japonica rice varieties. Genet Mol Biol 37(4):652–661

Yang J, Qu Y, Chen Q, Tang J, Lübberstedt T, Li H, Liu Z (2019) Genetic dissection of haploid male fertility in maize (Zea mays L.). Plant Breed 138:259–265

Yao L, Zhang Y, Liu C, Liu Y, Wang Y, Liang D, Kelliher T (2018) OsMATL mutation induces haploid seed formation in indica rice. Nat Plants 4(8):530–533

Yu W, Birchler JA (2015) A green fluorescent protein-engineered haploid inducer line facilitates haploid mutant screens and doubled haploid breeding in maize. Mol Breed 36:5

Zabirova ER, Chumak MV, Shatskaia OA, Scherbak VS (1996) Technology of the mass accelerated production of homozygous lines. Kukuruza Sorgo N 4:17–19

Zhang ZH (1989) The practicability of anther culture breeding. In: Mujeeb-Kazi A, Sitch LA (eds) Review of advances in plant biotechnology 1985–88: 2nd international symposium on genetic manipulation in crops. International Maize and Wheat Improvement Center/International Ricc Research Institute, Méx/Los Baños, pp 31–42

Zhang T, Talbert PB, Zhang W, Wu Y, Yang Z, Henikoff JG, Jiang J (2013) The CentO satellite confers translational and rotational phasing on cenH3 nucleosomes in rice centromeres. Proc Natl Acad Sci 110(50):E4875–E4883

Zhang Z, Liu Z, Hu Y, Li W, Fu Z, Ding D, Tang J (2014) QTL analysis of kernel-related traits in maize using an immortalized F2 population. PLoS One 9(2):e89645

Zhao YD, Hull AK, Gupta NR, Goss KA, Alonso J, Ecker JR, Normanly J, Chory J, Celenza JL (2002) (Trp)-dependent auxin biosynthesis in Arabidopsis: involvement of cytochrome P450s CYP79B2 and CYP79B3. Genes Dev 16:3100–3112

Zheng MY, Weng Y, Sahibzada R, Konzak CF (2003) Isolated microspore culture in maize (Zea mays L.), production of doubled-haploids via induced androgenesis. In: Maluszynski M, Kasha KJ, Forster BP, Szarejko I (eds) Doubled haploid production in crop plants: a manual. Springer, Dordrecht, pp 95–102

Zhou CHANG, Yang HY (1980) *In vitro* induction of haploid plantlets from unpollinated young ovaries of Oryza sativa. Acta Genet Sin 7(3):287–288

Zhou C, Yang HY (1981) *In vitro* embryogenesis in unfertilized embryo sacs of Oryza sativa L. Acta Bot Sin 23:176–180

Zhou C, Yang HY (1991) *In vitro* production of haploids in rice through ovary culture. In: Bajaj YPS (ed) Rice. Biotechnology in agriculture and forestry, vol 14. Springer, Berlin/Heidelberg, pp 180–192

Success Stories of Doubled Haploids

6

Suman Rawte, Garima Diwan, Suprasanna Penna,
and Zenu Jha

Abstract

The successful integration of doubled haploids in biotechnology and breeding has ushered in a new era of innovation in crop improvement. This chapter deals with the notable successful examples on the utilization of doubled haploids. Doubled haploids have become invaluable tools for crop breeders for the ability to rapidly generate homozygous lines. This approach has been particularly fruitful in enhancing crop traits such as disease resistance, yield, and stress tolerance, thus addressing critical challenges in global agriculture. Moreover, the use of doubled haploids has played a pivotal role in biotechnological advancements, enabling efficient gene mapping and the development of genetically modified crops. This technique, often employed in crops like wheat, maize, rice, brassica, chickpea cucumber, and apricot, has yielded prominent varieties with enhanced traits. Several superior rice varieties have been bred through haploid culture in China, Korea, Japan, and the United States. In wheat, varieties like DH-1 line have emerged as high-yielding cultivars with improved disease resistance. Barley has witnessed the development of varieties such as "Igri" and "Golden Promise," known for their enhanced malting quality and disease tolerance. It also highlights the transformative impact of doubled haploids on modern agriculture and emphasizes their importance in addressing the ever-growing global demand for food security and sustainable crop production.

S. Rawte · G. Diwan · Z. Jha (✉)
Department of Genetics and Plant Breeding, Indira Gandhi Krishi Vishwavidyalaya, Raipur, Chhattisgarh, India

S. Penna
Amity Center for Nuclear Biotechnology, Amity Institute of Biotechnology, Amity University Maharashtra, Mumbai, Maharashtra, India
e-mail: psuprasanna@mum.amity.edu

Z. Jha et al. (eds.), *Doubled Haploids: Technological Advances and Role In Crop Improvement*, https://doi.org/10.1007/978-981-96-2339-6_6

Keywords

Food Security · Doubled Haploids · Crop Plants · DH Cultivars · Crop Improvement

6.1 Introduction

Ensuring food security is decisive in addressing food shortages and providing for the global population. Consequently, it is imperative to enhance the precision and efficiency of agricultural scientific research. Employing advanced, rapid techniques can greatly expedite the development of crop varieties. Within the dominion of agricultural research, the utilization of haploid and doubled haploid plant regeneration through anther culture has proven to yield significant outcomes. Numerous scientists and researchers have successfully introduced new crop varieties using this approach.

Doubled haploid techniques are an important tool in plant breeding and genetics research. The production of haploids and DHs stands out as a highly advantageous biotechnological tool, and the establishment of haploidy technology and corresponding protocols for generating homozygous plants has had a substantial impact on agricultural systems. In contemporary agriculture, these biotechnologies have become essential components of breeding programs for numerous agriculturally significant crops.

Various methods exist for obtaining haploids and DHs, via anther culture, microspore culture, ovary and ovule culture, in vitro induction of haploids, genetic transformation, doubled haploid lines via mitotic chromosome doubling, genomic tools, and marker-assisted selection, among which in vitro anther or isolated microspore culture being the most efficient and widely employed approach.

One of the key advantages of doubled haploid crops is the reduction in time required for breeding cycles. Traditional breeding methods may take several years to achieve homozygosity through repeated cycles of self-pollination and selection. With doubled haploid technology, this process is dramatically expedited, potentially reducing the breeding timeline by half or more. This acceleration is particularly crucial in addressing global challenges such as food security and climate change, as it allows for the quicker development and deployment of crop varieties that are better adapted to changing environmental conditions, thereby contributing to more resilient and productive agricultural systems. The successful integration of doubled haploid technology into crop breeding programs marks a significant milestone in agricultural innovation, offering a promising avenue for addressing the complex challenges faced by modern agriculture. This chapter delves into the current state of knowledge regarding the success stories of production of haploids and DHs through different methods in different crops.

6.2 Historical Development of Haploids and Doubled Haploids

The concept of haploids, organisms with a single set of chromosomes, was first introduced in the early 1900s by Danish botanist Albert Blakeslee. He observed naturally occurring haploids in certain plants, identifying their potential for genetic research and breeding. In the 1920s, numerous researchers independently stumbled upon androgenesis, a phenomenon enabling the development of haploid plants from the pollen grains of specific species. This discovery laid the foundation for the creation of doubled haploids (Blakeslee et al. 1922). The refinement of tissue culture techniques during the 1970s and 1980s greatly streamlined the production of doubled haploids. Researchers could now efficiently culture plant tissues and manipulate them to induce and rogenesis.

The first agricultural crop used for haploid plant regeneration was *Hordeum vulgare* L., achieved through interspecific wide crossing. This method involved a combination of 28 chromosomes from the auto-tetraploid cultivated barley *Hordeum vulgare* and the tetraploid barley *H. bulbosum*. Selective chromosome elimination occurred reciprocally, as these chromosomes could not undergo crossing over. Eventually, this process yielded new double monoploid plantlets with 14 chromosomes, which proved to be fertile homozygotes (Kasha and Kao 1970). Subsequently, numerous researchers have directed their attention toward this strategy for the development of new and enhanced crop varieties. Over time, the utilization of doubled haploids extended to other vital crop plants, including maize, rice, barley, and various vegetable crops. This accelerated the breeding process and facilitated the creation of new and enhanced varieties (Table 6.1).

Today, doubled haploids continue to play a pivotal role in plant breeding and genetic research. They have revolutionized the development of new crop varieties by enhancing traits, shortening breeding cycles, and ensuring genetic purity. These advances are indispensable in addressing global agricultural challenges, such as bolstering crop yields and fortifying plants against pests, diseases, and environmental stressors.

6.3 Success Stories

The success stories of doubled haploids in plant breeding represent a significant breakthrough in agricultural science and crop improvement. Doubled haploids, with their ability to accelerate the development of pure, homozygous plant lines, have revolutionized the field of plant breeding. They have contributed to the creation of crop varieties with enhanced traits, including disease resistance, increased yield, and adaptability to changing environmental conditions. These success stories highlight the transformative impact of doubled haploids in modern agriculture and their crucial role in addressing global challenges related to food security and sustainable crop production.

Table 6.1 Development of DH new varieties in crops

S. no.	Crop	New varieties developed	Trait improvement	References
1	Rice	Huayu I, Huayu II, Xin Xiu, Late, Keng 959, Tunghua 1, Tunghua 2, Tunghua 3, Zhonghua 8, Zhonghua 9, Huahanzao, Huajian 7902, Tanghuo 2, Shanhua 7706, Huahanzao 77001, Nanhua 5, Noll, Hua 03, Hwacheongbyeo, Joryeongbyeo, and Hwajinbyeo	Exceptional grain quality, along with resistance to blast and bacterial blight diseases, early maturity, good quality and disease resistance, resistance to lodging, resistance to brown planthopper and stripe virus disease	Hu and Zeng (1984), Chen (1986), Loo and Xu (1986), Yang and Fu (1989), Lee et al. (1989), Mishra and Rao (2016)
2	Wheat	DH 114 (Him Pratham), Jinghua No-1, Florin, GK Délibáb, McKenzi, AC Andrew, GK Feh'er, DH-1 line, LH82, W438, and BSSS53	Robust resistance against 17 races of brown rust and 8 races of yellow rust, resistance to leaf blight	Chaudhary et al. (2013), Hu et al. (1986), De Buyser et al. (1987), Pauk et al. (1995), Graf et al. (2003), Sadasivaiah et al. (2004), Lantos et al. (2016), and Patial et al. (2021a)
3	Barley	Golden Promise, Igri, Barke, Compass, Flinders, Scope, and Spartacus	High malt quality, making them ideal for brewing	Schreiber et al. (2020) and Han et al. (2021)
4	Chickpea	CDC Xena, CDC Luna (kabuli), Sonali (desi), DH Elite, and DH Power	Early maturity, high yield potential, resistance to *ascochyta* blight and *fusarium* wilt, drought tolerance, and disease resistance	Grewal et al. (2009), Kumar and Chaudhary (2016), and Dwivedi et al. (2018)
5	Soybean	Anjasmoro and DH Speedy	Pod quality, pest resistance, and adaptability to different environments	Sumarmi et al. (2020) and Garda et al. (2020)
6	*Brassica* spp.	Ningyou7, Tapidor, Varuna and DH 1, 02-12-1, 87-534-3, C59 and CR9, Aiyuan1, Zhongyou 821, DS-4, SW Hickory, Early Big, Rapidase, and Gobi	Antioxidant properties, well-characterized genetic background, improved varieties	Shah et al. (2018), Zhao et al. (2019), Żur et al. (2022), Wei et al. (2021), Sharma et al. (2020), Alan et al. (2021), and Starosta et al. (2023)

(continued)

Table 6.1 (continued)

S. no.	Crop	New varieties developed	Trait improvement	References
7	Onion	Bhima Shubhra, SYN-YIX E, SYN-YIX C, BGS 151, VAZ 3, PLDR 17, and UDH066631	Improved onion varieties for traits such as disease resistance, yield, and bulb quality	Hyde et al. (2012) and Anandhan et al. (2014)
8	Tomato	Micro-Tom, Ohio 98114 DH, and UC82B DH	Compact size, rapid growth, and ease of cultivation, insect resistance	Niazian et al. (2009) and Zhao et al. (2014)
9	Cucumber	Borszczagowski, Gy14 DH, WI7120 DH, and H19 D	Disease resistance, fruit quality, and yield	Li et al. (2013) and Dong et al. (2016)
10	Sunflower	G11/1440, G11/1485, ANN1811, HA 60, and HA 89	Varietal development, enhance genetic diversity, improved varieties	Wang et al. (2023)

6.3.1 Cereal Crops

Promisingly, researchers Niizeki and Oono achieved the successful development of haploid regeneration techniques for *japonica* rice crops in 1968 (Niizeki and Oono 1968). In China high-yielding rice varieties via doubled haploid technology like Huayu I, Huayu II, Xin Xiu, Late, Keng 959, Tunghua 1, Tunghua 2, Tunghua 3, Zhonghua 8, Zhonghua 9, Huahanzao, Huajian 7902, Tanghuo 2, Shanhua 7706, Huahanzao 77001, Nanhua 5, Noll, and Hua 03 featuring exceptional grain quality, along with resistance to blast and bacterial blight diseases, was released by several scientists (Hu and Zeng 1984; Chen 1986; Loo and Xu 1986; Yang and Fu 1989). Some varieties were also developed in China via anther culture for traits like early maturity; good quality and disease resistance, resistant to lodging, resistance to brown planthopper and stripe virus disease (Mishra and Rao 2016).

Rice breeding programs aim to achieve several key objectives, including the development of rice varieties that exhibit enhanced yield, superior grain quality, and robust resistance to both abiotic and biotic stresses in addition to which the utilization of anther culture for generating doubled haploids has significantly expedited bio-fortification initiatives and hastened the introduction of some rice varieties. This technique has resulted in the release of numerous superior rice varieties, particularly the *japonica* type, in countries such as China, Korea, Japan, and the United States (Tripathy 2018). In India, several DH varieties have been developed including, Parag-401, Risabell, Janka, Abel, CR Dhan 10 (CRAC2221-43), Satyakrishna, CR Dhan 801 (CRAC2224-1041, IET 18720), Phalguni (Mishra and Rao 2016).

In South Korea anther-derived varieties of rice, namely, Hwacheongbyeo, Joryeongbyeo, and Hwajinbyeo, were developed for resistance to rice stripe *Tenuivirus*, brown planthopper, bacterial blight, and blast (Lee et al. 1989). Haploid and doubled haploid plantlets from salt-tolerant *indica* rice cultivars, namely, Pokkali and Nona Bokra, were successfully generated through anther culture under controlled in vitro conditions. These genotypes hold a great promise as valuable resources for future rice breeding (Javed et al. 2007). Resistance to iron chlorosis, high milling and cooking with superior grain quality, blast resistance, and drought and cold tolerance at an early stage were reported (Pauk et al. 2009). In the case of rice, a remarkable feature is that approximately 30–40% of the regenerants produced through anther culture naturally become doubled haploids, due to the spontaneous doubling of haploid cells (Mishra et al. 2013).

The combination of anther culture and molecular marker-assisted selection has empowered the creation of versatile breeding assets, including cytoplasmic male sterile maintainer lines (Wang et al. 2020). *Indica*-type rice suffer from poor callus induction and morphogenesis during callus induction and regeneration (Rahman et al. 2021); therefore, the constrained retrieval of doubled haploids has hampered the overall efficiency of androgenic methods, particularly in the context of *indica* rice (Mayakaduwa and Silva 2023). The expression analysis focuses on recent advancements in in vivo haploid induction protocols observed in other cereal crops and assesses their applicability to rice. Specifically, it delves into the present state of knowledge regarding in vivo haploid inducer systems employing *MTL* and *CENH3* mutants which has unveiled new prospects for the creation of in vivo haploid induction systems. The primary method for generating knockout mutants of rice's *MTL* and *CENH3* orthologues currently relies on the genome editing tool CRISPR/Cas9 (Mayakaduwa and Silva 2023).

In wheat (*Triticum aestivum* L.), by employing the wheat × maize method for doubled haploid induction, Chaudhary et al. (2013) achieved success in the creation of DH 114 (Him Pratham), marking the first doubled haploid wheat in the hill regions of the northwestern Himalayas in India. The production of doubled haploids is achieved through the utilization of alternative methods such as the one described by Castillo et al. in 2015 and the isolation of microspore cultures (Lantos and Pauk 2016). In the context of winter wheat, in vitro anther culture has proven to be a successful technique in various research initiatives aimed at introducing new wheat varieties. Prominent examples include Jinghua No-1 (Hu et al. 1986), Florin (De Buyser et al. 1987), GK Délibáb (Pauk et al. 1995), McKenzi (Graf et al. 2003), and AC Andrew (Sadasivaiah et al. 2004). The wheat × maize wide cross approach for producing doubled haploid plants is regarded as being two to three times more efficient in generating healthy green plantlets compared to anther culture. Numerous studies have illustrated the achievement of producing doubled haploid (DH) plants by utilizing maize pollen on bread wheat, as documented in reports by Amrani et al. (1993) and Niroula et al. (2007).

A spelt wheat cultivar, GK Feh'er, exhibited a highly efficient anther culture protocol, as reported by Lantos et al. (2016). The protocol resulted in a significant presence of embryo-like structures, reaching 62.2 per 100 anthers. Within this population, successful regeneration led to 30.6 green plantlets per 100 anthers. Among the regenerated plants, 89.0% displayed a green phenotype, while albino plants were scarce, with only 3.8 per 100 anthers. The latest and detailed anther culture procedure has been recently published by the same research group (Lantos and Pauk 2021).

Utilizing wheat F_1 hybrids (HS 542/China 84–40,022) crossed with *I. cylindrica*, Patial et al. (2021a) successfully generated the DH-1 line, exhibiting robust resistance against 17 races of brown rust and 8 races of yellow rust. Subsequently, these researchers conducted recent experiments to refine the conditions for DH wheat production involving *I. cylindrica* pollination (Patial et al. 2021b). The report outlines the optimization of (a) spike treatment with auxins post-pollination, (b) cultivation duration before embryo isolation, (c) the most suitable medium for culture, and (d) the colchicine concentration for haploid plant treatment. The findings revealed the optimal conditions as follows: (a) 50 mg/L 2,4-D + 50 mg/L dicamba, (b) 24 days post-pollination, (c) 1/2MS medium supplemented with 1 mg/L IAA, and (d) 1 mg/L kinetin and 0.075–0.08% colchicine, respectively.

In maize both in vitro and in vivo techniques can be employed to generate haploids. Nonetheless, in the case of in vitro methods, there is limited evidence of their ability to consistently yield the substantial quantities of doubled haploid lines essential for maize breeding programs, which is why they are presently not in use (Liu et al. 2016). The use of doubled haploids cuts the breeding cycle by half, allowing for the development of new varieties within just a few years. Successful implementation of DH technology in maize breeding programs depends on the ability to produce haploids. This is achieved by using pollen from haploid inducers to pollinate the source germplasm from which DH lines will be developed (Chaikam et al. 2019).

The newly created maize varieties displayed enhanced resistance to common maize pathogens, reducing crop losses significantly. Farmers who adopted these disease-resistant varieties experienced a substantial increase in their maize yields, leading to improved food security and income. LH82 is a doubled haploid maize line known for its resistance to leaf blight and other diseases, making it a valuable genetic resource for disease-resistant maize breeding. W438 is a maize inbred line that has been used in research to understand various traits related to maize development, including flowering time and disease resistance. The research team successfully developed doubled haploid maize lines with elevated protein content and improved amino acid profiles. BSSS53 is a doubled haploid maize line that has been used in breeding programs to improve the grain quality and nutritional content of maize.

Through doubled haploid technology, the research team achieved barley lines with consistently high malt quality, making them ideal for brewing. Golden Promise, Igri, and Barke are widely used barley doubled haploid lines known for their good malting quality. It has been used in breeding programs aimed at enhancing the malting and brewing qualities of barley varieties (Schreiber et al. 2020). Gene editing using CRISPR method was done in some commercial varieties of barley by developing an anther culture-based system that permits the effective creation of transgenic and gene-edited plants from commercial barley varieties like Golden Promise and four Australian varieties, namely, Compass, Flinders, Scope, and Spartacus (Han et al. 2021).

6.3.2 Pulse Crops

Doubled haploids have played a significant role in improving pulse crops, such as beans, peas, and lentils. These success stories are primarily related to faster breeding and the development of pure, homozygous lines. Chickpea (*Cicer arietinum* L.) was effectively cultivated using anther culture. To initiate the development of chickpea embryos, the researchers initially employed specific stress treatments, including exposure to cold stress, electric shock, centrifugation, and high osmotic pressure. Subsequently, electroporation treatment was utilized to promote root formation during the plantlet culture stage. Grewal et al. in 2009 developed doubled haploid using the Canadian cultivar CDC Xena, CDC Luna (kabuli), and the Australian cultivar Sonali (desi); it exhibits early maturity, high yield potential, and resistance to diseases such as *ascochyta* blight and *Fusarium* wilt. The article authored by Croser et al. in 2006 provides valuable insights into various research strategies, advancements, challenges, and prospects related to the production of doubled haploids in the Fabaceae family. Kumar and Chaudhary (2016) developed a doubled haploid by microspore culture named DH Elite, which facilitates the development of pure lines with enhanced traits, including disease resistance and yield, in common bean breeding programs.

A DH variety, Chickpea DH Power, was developed to expedite chickpea breeding by producing homozygous lines with the desired traits like drought tolerance and disease resistance (Dwivedi et al. 2018). Anjasmoro soybean cultivar was developed to accelerate soybean breeding programs by producing true breeding lines in a single generation via microspore culture by Sumarmi et al. (2020). Similarly, Garda et al. (2020) focused on traits such as pod quality, pest resistance, and adaptability to different environments and developed DH Speedy variety of soybean.

6.3.3 Vegetable Crops

Doubled haploid technology has been used to develop disease-resistant and high-yielding varieties. For example, researchers have used this technique to create cucumber lines that are resistant to common diseases like downy mildew and cucumber mosaic virus. The resulting cucumber varieties are not only more robust but also have a more consistent fruit size and quality, making them popular choices for both commercial growers and home gardeners. Powdery mildew, caused by the fungus *Sphaerotheca fuliginea*, has been noted to affect melons, and this issue can become apparent even at the haploid stage (Kuzuya et al. 2003).

Investigations into gametic embryogenesis have been extended to various members of the Brassicaceae family. It involved the induction anther cultures and subsequently in cultures of isolated microspores from white cabbage (*Brassica oleracea* var. *capitata*), Brussels sprouts (*B. oleracea* var. *gemmifera*), cauliflower (*B. oleracea* var. *italica*), and radish (*Raphanus sativus*). The utilization of a desiccation process on embryos generated through abscisic acid (ABA) treatment resulted in a notable enhancement in the transformation of embryos into plants. First, the homozygosity, ploidy, and trait uniformity of the DH cabbage plants were examined. Afterward, the DH plants were self-pollinated to generate DH lines, which were then analyzed for their antioxidant properties (Żur et al. 2022). Tapidor, Varuna, and DH 1 are winter oilseed rape (*Brassica napus* L.) DH lines, which have been extensively used in genetic and genomic studies due to its well-characterized genetic background.

02-12-1 is a DH line of cabbage (*Brassica oleracea* var. *capitata*) developed for research and breeding purposes. DH lines like 02-12-1 are valuable resources in cabbage breeding programs for their uniformity and homozygosity, which accelerate the development of new cabbage varieties with the desired traits. 87-534-3 is another DH line of cabbage (*Brassica oleracea* var. *capitata*) developed for research and breeding purposes. Like other DH lines, it offers a uniform and homozygous genetic background, facilitating cabbage breeding efforts. C59 and CR9 are DH lines of cabbage (*Brassica oleracea* var. *capitata*) used in research and breeding programs. Its homozygous nature aids in the selection and development of improved cabbage varieties. These investigations highlight that by implementing laboratory-optimized protocols, it is feasible to acquire and distinguish valuable DH plants in the context of cabbage cultivation (Żur et al. 2022).

Since 1995, research into gynogenesis within the onion (*Allium cepa* L.) has been ongoing, driven by the necessity to expedite the breeding process for biennial crop species. One of the focal methods employed is the in vitro cultivation of the entire flower buds that gave a favorable outcome (Michalik et al. 1997). Subsequent investigations have revealed that the success of gynogenesis in onions significantly depends on the genetic makeup and the composition of the culture medium. Once the ideal parameters, including flower bud size, medium, light conditions, and regeneration conditions, were established, the result was the acquisition of haploid embryos. A significant accomplishment was the formulation of a protocol for the diploidization of gynogenic embryos. This is of paramount significance since haploids are sterile, and the spontaneous occurrence of diploidization in onions is a rare phenomenon (Bohanec 2002). Within the category of Indian short-day cultivars, it was observed that the gynogenic potential was most pronounced in Bhima Shubhra, as reported by Anandhan et al. in 2014. SYN-YIX E, SYN-YIX C, BGS 151, VAZ 3, PLDR 17, and CUDH066631 are some DH lines of onion (*Allium cepa*) developed for research and breeding purposes (Hyde et al. 2012). It has been used in genetic and genomic studies aimed at improving onion varieties for traits such as disease resistance, yield, and bulb quality.

In the context of tomato cultivation, investigations have explored the induction of haploidy through an alternative culture method. Nonetheless, the outcomes have not yet reached a level of satisfaction that would permit its regular integration into breeding programs. Despite this, considering the potential of the process of acquiring new lines through this technique, there is an interesting incentive to persist in the quest for alternatives that can efficiently produce doubled haploids in the tomato crop (Niazian et al. 2009). Gynogenic haploid induction stands as an alternative method to anther culture, with several researchers, such as Bal and Abak in 2007, presenting successful reports on gynogenesis in tomatoes. Micro-Tom, Ohio 98114

DH, and UC82B DH are some doubled haploid (DH) lines derived from the miniature tomato. It is particularly popular in research due to its compact size, rapid growth, and ease of cultivation, making it ideal for genetic and molecular studies. The future holds the promise of gynogenesis becoming a potential tool for haploidization in tomatoes, as suggested by Zhao et al. (2014).

Gynogenesis is considered the less preferred method for inducing haploids due to its low efficiency, and it is applicable only to species that do not respond well to more efficient techniques. Gynogenesis presents a promising path for the development of haploids in cucurbits (Dong et al. 2016). Borszczagowski, Gy14 DH, WI7120 DH, and H19 DH are doubled haploid (DH) lines of cucumber developed for research and breeding purposes. It offers a uniform genetic background, making it valuable for studying cucumber genetics and breeding for traits such as disease resistance, fruit quality, and yield. In the case of cucumber, there have been reports of successful unpollinated ovule culture using a cucumber basal medium enriched with thidiazuron and silver nitrate, as documented by Li et al. in 2013.

6.3.4 Oilseed Crops

In *B. napus*, the generation of haploid plants is typically accomplished through microspore culture, with rapeseed as the principal species for investigating microspore-derived plants. Anther culture, while considered a viable approach, is generally less efficient in producing haploid embryos in *B. napus*. It is a naturally occurring hybrid, resulting from the crosses of *Brassica rapa* and *Brassica oleracea*, followed by the process of chromosome doubling. This unique origin makes it a valuable model species for the study of genome duplication. Additionally, certain plant species have the capability to produce haploid embryos through the crossing of specific individuals with specific genetic traits.

A novel finding shed light on the molecular mechanisms responsible for the dwarfism observed in *B. napus*. Using "Aiyuan1" DH lines and the "Zhongyou 821" cultivar, Zhao et al. (2019) and Wei et al. (2021) successfully identified a newly discovered dwarf locus, DS-4, which encodes a nucleus-targeted auxin/indole-3-acetic acid (Aux/IAA) protein. A total of 36 genes linked to the regulation of flowering time were discovered using comprehensive transcriptome analysis of the doubled haploid semi-wintertime line known as "Ningyou7" (Shah et al. 2018). Incorporating doubled haploid *B. napus* lines has been demonstrated to be advantageous in transcriptomic and metabolomic investigations, yielding valuable insights into the initial responses to clubroot in both resistant and susceptible *B. napus* lines. This includes the detection of alterations in organic acid, amino acids, sugar profiles, and the accumulation of mannitol in roots. Additionally, variations in the expression of pyrroline-5-carboxylate, citrate synthase, and pyruvate kinase were also observed (Shaikh et al. 2020).

Microspore culture was done to develop doubled haploid lines from interspecific cross between *B. napas* and *B. rapa*. NR1 (also known as Tapidor × R-o-18) and KR1 (also known as Kikuchi × R-o-18) are some examples of doubled haploid assessing seed germination under drought conditions using PEG-6000 that highlighted the superior drought tolerance of doubled haploid plants from interspecific hybrids compared to their parent cultivars. Furthermore, the oil extracted from the seeds of these doubled haploid plants exhibited excellent nutritional quality (Daurova et al. 2020). The most effective approach for triggering parthenogenesis, using γ-irradiated pollen, is recognized as the primary method for generating haploid sunflower plants. Dihaploid lines produced via γ-irradiated pollen like SW Hickory of *B. napas*, Early Big of *B. oleracea*, Rapidase line of *B. rapa*, and Gobi DH line of *B. juncea* have been actively integrated into commercial breeding programs, resulting in the development of high-yielding F_1 hybrids equipped with heightened resistance to abiotic and/or biotic stresses (Sharma et al. 2020; Alan et al. 2021; Starosta et al. 2023).

Conventional in vitro methods to induce haploids in *Helianthus* plants, whether through male (isolated microspore and anther cultures) or female (unpollinated ovaries and ovules culture) gametophytes, remain underutilized in biotechnological applications due to their limited embryogenic response. A transformative shift in this scenario can be realized by identifying genotypes with a robust potential for gametic embryogenesis through a fine-mapping approach involving quantitative trait loci. The development of sunflower doubled haploids via gamma-induced parthenogenesis has enabled the breeding of lines with resistance to a range of diseases including downy mildew (*Plasmopara helianthi Novot*), phoma (*Phomamacdonaldii*), phomopsis (*Phomopsis helianthi*), *Alternaria* leaf spot (*Alternaria* spp.), and broomrape (*Orobanche cumana*) as well as herbicide tolerance, specifically imidazolinone resistance. Also anther culture was employed to yield highly fertile F_1 sunflower hybrids, which were subsequently utilized in heterosis breeding (Mabuza et al. 2023). The varieties G11/1440, G11/1485, ANN1811, HA 60, and HA 89 have undergone improvement through the doubled haploid method in sunflower breeding, resulting in enhanced traits and more stable genetic backgrounds for better performance in various agricultural works (Wang et al. 2023).

6.3.5 Fruits Crops

The significant promise of utilizing haploidy, doubled haploidy, and gametic embryogenesis in fruit breeding is clearly apparent. Haploids ($1x = 1n = 8$) identified in peach (*Prunus persica* (L.) Batsch) emerged naturally, displaying characteristic traits of haploidy such as slender stems, narrow leaves, limited vegetative growth, and smaller, less fertile flowers. Research conducted by Scorza and Pooler in 1999 regarding the growth and yield of F_1 hybrid peaches derived from doubled haploids (DH) revealed comparable productivity to standard cultivars. However, F_1 hybrids offered distinct advantages in producing consistent seedling scion cultivars. In the case of citrus, haploid seedlings were initially achieved through the application of gamma radiation. From a cross between a diploid *Clementine mandarin* and a diploid *Pearl tangelo*, a haploid embryo was obtained from an immature seed.

Additionally, the induction of in situ parthenogenesis through the irradiation of Meyer lemon (*Citrus meyeri* Y. Tan.) pollen at 300, 600, and 900 Gray (Gy) using a cobalt 60 source led to the production of nine haploid plantlets. Although they did not survive, two embryogenic callus lines in clementine (*Citrus clementina* Hort. Ex Tan.), and cv. SRA 63 were obtained; apart from this, some DH varieties such as Valencia, Navel, Meyer Lemon, and Ruby Red were also reported (Jain et al. 2022).

Four homozygous sweet cherry (*Prunus avium* L.) lines were successfully regenerated through in situ parthenogenesis, initiated by pollination with irradiated pollen, and subsequent embryo and cotyledon culture in the "Altenburger" cultivar, as documented in a study, and the gametic origin of the calluses in cherry was confirmed through isozyme analyses (Hofer and Grafe 2003). Peixe et al. (2004) documented the development of calluses on cultured anthers of the "Harcot" apricot, along with the formation of nodular structures. The most favorable outcomes were achieved with a heat pre-treatment at 28 °C for 8 days, surpassing the results obtained at 36 °C or 24 °C.

Olive stands as one of the essential fruit and paramount oil-producing plants native to the Mediterranean region. Olive is a diploid species ($2n = 2x = 46$), known for its outcrossing behavior and remarkable longevity. Notably, the isolated microspore culture of two olive cultivars, namely, Arbequina and Picual, has demonstrated successful induction of sporophytic division, multinucleate microspores, and the development of multicellular structures.

Due to the dioecious nature of the mulberry plant, inbreeding as a means to obtain haploids and homozygous plants is not a viable approach. However, gynogenic haploids from a female clone of mulberry (*Morus alba* L. Cv. K-2) were successfully obtained through in vitro culture of unpollinated ovaries derived from in vitro developed inflorescences. It's worth noting that attempts at anther culture for producing haploids in this particular tree crop have not yielded successful results. In the case of papaya (*Carica papaya*), which is a polygamous species, the primary breeding systems that rely on true-bred lines significantly benefit from haploid induction achieved via anther culture. Haploid plantlets and embryos derived from pollen were successfully obtained by culturing papaya anthers at the uninucleate stage. Co7, Pusa Nanha, and Maradol Roja DH lines were developed by Anandan et al. (2011), Gyanchand et al. (2015), and Posada-Pérez et al. (2017), respectively, for improvement of the cultivar. The induction of haploid plantlets through anther culture was accomplished in a growth medium devoid of any growth regulators, all while maintaining dark conditions.

Kiwifruit, specifically the Hayward cultivar, which is a hexaploid species ($2n = 6x = 174$), was subjected to irradiated pollen to induce the development of parthenogenetic tri-haploids. The most favorable outcomes were achieved with irradiation dosages ranging from 500 to 1500 Gy. It's important to note that the genetic makeup of the pollen parent had a significant impact on the success of obtaining both seedlings and tri-haploids.

Assani et al. (2003) reported the successful production of 41 haploid ($n = x = 11$) banana plants through anther culture of *Musa balbisiana* (BB). This achievement comprised 18 plants from the Pisang Klutuk genotype, 12 from Pisang Batu, 7 from

Pisang Klutuk Wulung, and 4 from Tani. Notably, the frequency of callus induction reached 77%, with approximately 8% of anthers giving rise to embryos after a 6-month cultivation period. The formation of embryos varied depending on the genotype, and the anther culture was conducted at a temperature of 27 °C under dark conditions. Haploids and doubled haploids have been generated through in situ parthenogenesis, induced by irradiated pollen, or via seedling selection from three pear cultivars: "Doyenne du Comice," "Williams," and "Harrow Sweet." The production of doubled haploids was achieved through either spontaneous chromosome doubling or treatment with oryzalin (200–300 μM). It's worth noting that pear anther culture of cv. Le Lectier resulted in the generation of two embryos, although their origin remained undetermined, and plant regeneration was not achieved (Kadota and Niimi 2004).

6.4 Conclusion

In the context of current challenges of food production and food security, it is imperative to advance DH technology for recalcitrant plant species that have previously shown limited success in haploid or doubled haploid plant regeneration. In the realm of crop improvement, the successful implementation of doubled haploid (DH) technology has undoubtedly propelled agricultural progress to new heights. Through the concerted efforts of researchers and breeders, coupled with advancements in biotechnology, the success of DH technology in crops, is exemplified by the varieties Huayu I, Huayu II, Xin Xiu, DH 114 (Him Pratham), Jinghua No-1, Florin, GK Délibáb, McKenzi, Golden Promise, Igri, Barke, Compass, Flinders, Scope, Spartacus, CDC Xena, CDC Luna (kabuli), Sonali (desi), DH Elite, Anjasmoro, DH Speedy, Zhongyou 821, DS-4, SW Hickory, Early Big, Rapidase, Gobi, Micro-Tom, Ohio 98114 DH, and UC82B DH in different crops, and this, stands as a testament to the transformative potential of genetic innovation in agriculture.

References

Alan AR, Celebi-Toprak F, Lachin A, Yildiz D, Gozen V, Besirli G (2021) Doubled haploid broccoli (Brassica olearacea var. italica) plants from anther culture. Methods Mol Biol (Clifton, NJ) 2288:201–216

Amrani N, Sarrafi A, Allibert G (1993) Genetic variability for haploid production in crosses between tetraploid and hexaploid wheats with maize. Plant Breed 110:123–128

Anandan R, Thirugnana KS, Sudhakar D, Balasubramanian P (2011) In vitro organogenesis and plantlet regeneration of (Carica papaya L.). J Agric Technol 7(5):1339–1348

Anandhan S, Chavan AA, Gopal J, Mote SR, Shelke PV, Lawande KE (2014) Variation in gynogenic potential for haploid induction in Indian short-day onions. Indian J Genet 74(2):526–528

Assani A, Bakry F, Kerbellec F, Haicour R, Wenzel G, Foroughi-Wehr B (2003) Production of haploids from anther culture of banana [Musa balbisiana (BB)]. Plant Cell Rep 21(6):511–516

Bal U, Abak K (2007) Haploidy in tomato (Lycopersicon esculentum Mill.): a critical review. Euphytica 158:1–9

Blakeslee AF, Belling J, Farnham ME, Bergner AD (1922) A haploid mutantin the jimson weed, "Datura stramonium". Science 55:646–647

Bohanec B (2002) Doubled-haploid onions. In: Rabinowitch HD, Currah L (eds) Allium crop science: recent advances. CABI Publishing, pp 145–157. https://doi.org/10.1079/9780851995106.0145

Castillo AM, Sánchez-Díaz RA, Vallés MP (2015) Effect of ovary induction on bread wheat anther culture: ovary genotype and developmental stage, and candidate gene association. Front Plant Sci 6:402

Chaikam V, Molenaar W, Melchinger AE, Boddupalli PM (2019) Doubled haploid technology for line development in maize: technical advances and prospects. Theor Appl Genet 132:3227–3243

Chaudhary HK, Tayeng T, Kaila V, Rather SA (2013) Enhancing the efficiency of wide hybridization mediated chromosome engineering for high precision crop improvement with special reference to wheat 9 Imperata cylindrica system. Nucleus 56(1):7–14

Chen Y (1986) The inheritance of rice pollen plant and its application in crop improvement. In: Hu H, Yang H (eds) Haploids of higher plants *in vitro*. Springer-Verlag, Berlin, pp 118–136

Croser JS, Lülsdorf MM, Davies PA, Clarke HJ, Bayliss KL (2006) Toward doubled haploid production in the Fabaceae: progress, constraints, and opportunities. Crit Rev Plant Sci 25:139–157

Daurova A, Daurov D, Volkov D, Zhapar K, Raimbek D, Shamekova M, Zhambakin K (2020) Doubled haploids of interspecific hybrids between Brassica napus and Brassica rapa for canola production with valuable breeding traits. OCL 27:45

De Buyser J, Henry Y, Lonnet P, Herzog R, Hespel A (1987) Florin – a doubled haploid wheat variety developed by the anther culture method. Plant Breed 98:53–56

Dong Y, Zhao W, Li X, Liu X, Gao N, Huang J, Wang W, Xu X, Tang Z (2016) Androgenesis, gynogenesis and parthenogenesis haploids in cucurbit species. Plant Cell Rep 35:1991–2019

Dwivedi SL, Siddique KH, Farooq M, Thornton PK, Ortiz R (2018) Using biotechnology-led approaches to uplift cereal and food legume yields in dryland environments. Front Plant Sci 9:393361

Garda M, Hale B, Rao N, Lowe M, Bright M, Goodling S, Phillips GC (2020) Soybean androgenesis I: identification of pyramidal stressors in anther cultures that sustain cell divisions and putative embryo formation from isolated microspore cultures. In Vitro Cell Dev Biol Plant 56:415–429

Graf RJ, Hucl P, Orshinsky BR, Kartha KK (2003) "McKenzi" hard red spring wheat. Can J Plant Sci 83:565–569

Grewal RK, Lulsdorf M, Croser J, Ochatt S, Vandenberg A, Warkentin TD (2009) Doubled-haploid production in chickpea (Cicer arietinum L.): role of stress treatments. Plant Cell Rep 28(8):1289–1299

Gyanchand, Sharma M, Kumar S, Sagar S, Kumar V, Chaurasia M (2015) In-vitro androgenesis in papaya (Carica papaya L.) cv. Pusa Nanha. J Appl Nat Sci 7:273–278

Han Y, Broughton S, Liu L, Zhang X, Zeng J, Xiaoyan H, Li C (2021) Highly efficient and genotype-independent barley gene editing based on anther culture. Plant Commun 2(2):100082

Hofer M, Grafe C (2003) Induction of doubled haploids in sweet cherry (Prunus avium L.). Euphytica 130(2):191–197

Hu H, Zeng JZ (1984) Development of new varieties via anther culture. In: Ammirato PV, Evans DA, Sharp WR, Yamada Y (eds) Handbook of plant cell culture: III. Crop species. Macmillan, New York, pp 65–90

Hu DF, Yuan ZD, Tang YL, Liu JP (1986) "Jinghua No-1" a winter-wheat variety derived from pollen sporophyte. Sci China Ser B – Chem Biol Agric Med Earth Sci 29:733–745

Hyde PT, Earle ED, Mutschler M (2012) Doubled haploid onion (Allium cepa L.) lines and their impact on hybrid performance. Hortic Sci 47:1690–1695

Jain S, Maurya P, Saini S, Jagga S, Jayachandran A, Kumar V, Kiran B (2022) Anther culture and double-haploid production in fruit crops: status & opportunities. J Pharm Innov 11(11S):975–985

Javed MA, Ishii T, Kamijima O, Misoo S (2007) The role of alternating culture temperatures and maltose in enhancing the anther culture efficiency of salt tolerant indica rice (*Oryza sativa* L.) cultivars, Pokkali and Nona Bokra. Plant Biotechnol 24:283–287

Kadota M, Niimi Y (2004) Production of triploid plants of Japanese pear (Pyrus pyrifolia Nakai) by anther culture. Euphytica 138(2):141–147

Kasha KJ, Kao KN (1970) High frequency haploid production in barley (*Hordeum vulgare* L.). Nature 225:874

Kumar G, Chaudhary N (2016) Induced cytomictic variations and syncyte formation during microsporogenesis in *Phaseolus vulgaris* L. Cytol Genet 50:121–127

Kuzuya M, Hosoya K, Yashiro K, Tomita K, Ezura H (2003) Powdery mildew (Sphaerotheca fuliginea) resistance in melon is selectable at the haploid level. J Exp Bot 54:1069–1074

Lantos C, Pauk J (2016) Anther culture as an effective tool in winter wheat (Triticum aestivum L.) breeding. Russ J Genet 52:794–801

Lantos C, Pauk J (2021) *In vitro* anther culture for doubled haploid plant production in spelt wheat. In: Segui-Simarro JM (ed) Doubled haploid technology. Methods in molecular biology, vol 2287. Humana, New York, pp 257–266

Lantos C, Jenes B, Bóna L, Cserhati M, Pauk J (2016) High frequency of doubled haploid plant production in spelt wheat. Acta Biol Cracov Ser Bot 58. https://doi.org/10.1515/abcsb-2016-0014

Lee YT, Lim MS, Kim HS, Shin HT, Kim CH, Bae SH, Cho CI (1989) An anther derived new high-quality variety with disease and insect resistance "Hwacheongbyeo". Res Rep Rural Dev Admin Rice 31(2):27–34

Li JW, Si SW, Chang JY (2013) Thidiazuron and silver nitrate enhanced gynogenesis of unfertilized ovule cultures of Cucumis sativu. Biol Plant 57:164–168

Liu Z, Wang Y, Ren J, Mei M, Frei UK, Trampe B, Lübberstedt T (2016) Maize doubled haploids. Plant Breed Rev 40:123–166

Loo SW, Xu ZH (1986) Rice: anther culture for rice improvement in China. In: Bajaj YPS (ed) Biotechnology in agriculture and forestry: II. Crops I. Springer, Berlin, pp 139–156

Mabuza LM, Mchunu NP, Crampton BG, Swanevelder DZH (2023) Accelerated breeding for Helianthus annuus (sunflower) through doubled haploidy: an insight on past and future prospects in the era of genome editing. Plan Theory 12:485

Mayakaduwa R, Silva T (2023) Haploid induction in indica rice: exploring new opportunities. Plan Theory 12:3118. https://doi.org/10.3390/plants12173118

Michalik B, Adamus A, Nowak E (1997) Induction of haploid plants in Polish onion cultivars. Acta Hortic 447:377–378

Mishra R, Rao GJN (2016) In-vitro androgenesis in rice: advantages, constraints and future prospects. Rice Sci 23(2):57–68

Mishra R, Rao GJN, Rao RN (2013) Effect of cold pretreatment and phytohormones on anther culture efficiency of two indica rice (Oryza sativa L.) hybrids: Ajay and Rajalaxmi. J Exp Biol Agric Sci 1(2):69–76

Niazian M, Shariatpanahi ME, Abdipour M, Oroojloo M (2009) Modeling callus induction and regeneration in an anther culture of tomato (Lycopersicon esculentum L.) using image processing and artificial neural network method. Protoplasma 256:1317–1332

Niizeki H, Oono K (1968) Induction of haploid rice plant from anther culture. Proc Jpn Acad 44(6):554–557

Niroula RK, Thapa DB, Bimb HP, Sah BP, Nayak S (2007) Production of haploid wheat plant from wheat (Triticum aestivum L.) × maize (Zea mays L.) cross system. Himalayan J Sci 5:47–52

Patial M, Chaudhary HK, Sharma N, Gangwar OP, Kishore N, Pal D, Pramanick KK, Bhardwaj SC, Chauhan R (2021a) Developing genetic stock for yellow and brown rust resistance in Triticum aestivum L. via Imperata cylindrica mediated doubled haploidy technique. Cereal Res Commun 49:1–10. https://doi.org/10.1007/s42976-021-00180

Patial M, Chaudhary HK, Sharma N, Sundaresha S, Kapoor R, Pal D, Pramanick KK, Shukla AK, Kumar J (2021b) Effect of different *in vitro* and in vivo variables on the efficiency of doubled haploid production in Triticum aestivum L. using Imperata cylindrica-mediated chro-

mosome elimination technique. Cereal Res Commun 49:133–140. https://doi.org/10.1007/s42976-020-00069-2
Pauk J, Kertész Z, Beke B, Bóna L, Csősz M, Matuz J (1995) New winter wheat variety – "GK Délibáb" developed via combining conventional breeding and invitro and rogenesis. Cereal Res Commun 23:251–256
Pauk J, Janeso M, Simon-Kiss I (2009) Rice doubled haploids and breeding. In: Touraev A, Forster BP, Jain SM (eds) Advances in haploid production in higher plants. Springer, Netherlands, pp 189–197
Peixe A, Barroso J, Potes A, Pais MS (2004) Induction of haploid morphogenic calluses from *in vitro* cultured anthers of Prunus armeniaca cv. "Harcot". Plant Cell Tissue Organ Cult 77(1):35–41
Posada-Pérez L, Montesinos YP, Guerra DG, Daniels D, Gómez-Kosky D (2017) Complete germination of papaya (Carica papaya L. cv. "Maradol Roja") somatic embryos using temporary immersion system type RITA® and phloroglucinol in semi-solid culture medium. In Vitro Cell Dev Biol Plant 53:505–513
Rahman ZA, Seman ZA, Othman AN, Ghaffar MBA, Razak AB, Yusof MFM (2021) Efficient callus induction and plant regeneration of Malaysian indica rice MR219 using anther culture. Biocatal Agric Biotechnol 31:101865
Sadasivaiah RS, Perkovic SM, Pearson C, Postman B, Beres BL (2004) Registration of "AC Andrew" wheat. Crop Sci 44:696–697
Schreiber M, Mascher M, Wright J, Padmarasu S, Himmelbach A, Heavens D, Milne L, Clavijo B, Stein N, Waugh R (2020) A genome assembly of the barley "transformation reference" cultivar golden promise. G3-Genes Genom Genet 10:g3.401010.2019. https://doi.org/10.1534/g3.119.401010
Scorza R, Pooler MR (1999) Growth and yield of F_1 Hybrid peaches developed from doubled haploids. Hortic Sci 34(5):928–931
Shah S, Weinholdt C, Jedrusik N, Molina C, Zou J, Große I, Schiessl S, Jung C, Emrani N (2018) Whole-transcriptome analysis reveals genetic factors underlying flowering time regulation in rapeseed (Brassica napus L.). Plant Cell Environ 41:1935–1947
Shaikh R, Farid M, Rahman H (2020) Inheritance of resistance to the newly identified Plasmodiophora brassicae Pathotypes in Brassica napus L. Can J Plant Pathol 43:256266
Sharma R, Chaoudhary P, Agarwal A (2020) Doubled haploid production in Brassica oleracea L.: a review. Plant Arch 20:6293–6300
Starosta E, Szwarc J, Niemann J, Szewczyk K, Weigt D (2023) Brassica napus haploid and double haploid production and its latest applications. Curr Issues Mol Biol 45(5):4431–4450
Sumarmi S, Daryono BS, Rachmawati D, Indrianto A (2020) Determination of soybean (*Glycine max* L. [Merrill]) microspores development stage based on the length of flower buds. Berk Penelit Hayati 20(1):6–11
Tripathy SK (2018) Anther culture for double haploid breeding in rice-a way forward. Rice Genom Genet 9:1–6
Wang P, Bai YL, Wang MX, Hu BH, Pu ZG, Zhang ZY, Zhang Q, Xu DW, Luo WL, Chen ZQ (2020) Breeding of CMS maintainer lines through anther culture assisted by high-resolution melting-based markers. J Integr Agric 19:2965–2973
Wang H, Hou H, Jan C-C, Chao WS (2023) Irradiated pollen-induced parthenogenesis for doubled haploid production in sunflowers (Helianthus spp.). Plants 12(13):2430
Wei T, Zhang L, Zhu R, Jiang X, Yue C, Su Y, Ren H, Wang MA (2021) Gain-of-function mutant of Iaa7 inhibits stem elongation by transcriptional repression of Expa5 genes in Brassica napus. Int J Mol Sci 22:9018
Yang XR, Fu HH (1989) Hua-03: a high protein indica rice. Intl Rice Res Nwsl 14(3):14–15
Zhao H, Wang X, Yong-chen D, De-wei Z, Yan-mei G, Jianchang G, Fei L, Snyder JC (2014) Haploid induction via *in vitro* Gynogenesis in tomato (Solanum lycopersicum L.). J Integr Agric 3119(13):60672–60673

Zhao B, Wang B, Li Z, Guo T, Zhao J, Guan Z, Liu K (2019) Identification and characterization of a new dwarf locus DS-4 encoding an aux/IAA7 protein in Brassica napus. Theor Appl Genet 132:1435–1449

Żur I, Adamus A, Cegielska TT, Cichorz S, Dubas E, Gajecka M, Juzoń SK, Kielkowska A, Malicka M, Oleszczuk S, Skrzypek E, Szala L, Szarejko I (2022) Doubled haploids: contributions of Poland's academies in recognizing the mechanism of gametophyte cell reprogramming and their utilization in breeding of agricultural and vegetable species. Acta Soc Bot Pol 91:1–32

The manufacturer's authorised representative in the EU is Springer Nature Customer Service Centre GmbH, Europaplatz 3, 69115 Heidelberg, Germany. If you have any concerns regarding our products, please contact ProductSafety@springernature.com

Printed and bound by CPI Group (UK) Ltd, Croydon, CR0 4YY
30/09/2026
02225302-0002